MANAGING THE FLOW OF TECHNOLOGY

MANAGING THE FLOW OF TECHNOLOGY: TECHNOLOGY TRANSFER AND THE DISSEMINATION OF TECHNOLOGICAL INFORMATION WITHIN THE R&D ORGANIZATION

Thomas J. Allen

This book was set in IBM Composer Baskerville by Vermont Book Services,
printed on R&E Book by Halliday Lithograph Corp., and bound in Holliston
Roxite B-51531 by Halliday Lithograph Corp., in the United States of
America.

Library of Congress Cataloging in Publication Data

Allen, Thomas John, 1931–
  Managing the flow of technology.

  Bibliography: p.
  Includes index.
  1. Technology transfer. 2. Communication of technical information.
3. Research, Industrial—Management. I. Title.
T174.3A44   658.4'5   76-57670
ISBN 0-262-01048-8

To My Family

CONTENTS

ACKNOWLEDGMENTS

In a sense, this is the biography of a research program. It is a report on a program of research that has been under way for more than a decade at MIT's Sloan School of Management. In fact, it is an interim report, since the research is continuing in a number of directions.

Many of the results reported in this volume have already appeared in journal articles. In my view and in the eyes of officials at the Office of Science Information Service, National Science Foundation, however, it seemed to be appropriate to integrate all of these results in a single volume.

The research began in 1963 when I, then a research engineer for the Boeing Company, inquired why the Management School at MIT, which taught courses in marketing, finance, production management, and so on, did not have a course in the management of research and development. That simple question, directed at the late Professor Donald G. Marquis, set into operation a series of events that have culminated in this volume.

Professor Don Marquis was an unusual man, a man of enormous talent, which is reflected in the many accomplishments of a long career spanning several fields. Don was also a gambler; he gambled on people. His response to the inquiry was simple and direct. He replied that R&D management was an area generally neglected by business schools because so little was known about it. At that particular time, however, he was just beginning to organize a research effort under the sponsorship of the National Aeronautics and Space Administration to fill this void. He then challenged me to join his team, suggesting a leave of absence from Boeing to work with this program of "research on research."

This was certainly a gamble—a long shot at best. I was completely untrained and inexperienced in the social sciences. The only asset I brought with me was an interest in the problem. Now, after many years of effort, perhaps Don's gamble has paid off. The reader, of course, will determine the value of this payoff.

Of course no research effort is possible these days without the contribution of many minds and hands. A key element in all of this was Don Marquis. He offered the initial stimulus and continuing encouragement up to his death in 1973. Next in importance is Professor Ed Roberts of the Sloan School, whose intellectual support and encouragement have been crucial over the years. Then there are the many graduate students, who have contributed so much to the data collection and analysis. Because there is a built-in turnover rate in the university, this list is necessarily long.

The first research assistants were J. Randall Brown and Maurice Andrien. They were followed by Richard J. Bjelland, Stephen I. Cohen, Daniel S. Frischmuth, Arthur Gerstenfeld, Paul W. O'Gara, Peter F. Gerstberger, William M. Collins, Robert L. Gipson, Grenville V. Craig, Arunkumar H. Firodia, Mario Y. Muñoz, James M. Piepmeier, Jehangir Mistry, Maurice Stouffer, Thaddeus W. Usowicz, Alan Fusfeld, and William McCarter.

The Office of Science Information Service of the National Science Foundation has generously supported this research since nearly its beginning in 1963. The members of that office, particularly Jack Scopino, Don Pollock, and Ed Weiss, have been extremely helpful in advising and encouraging the research program. Their role has encompassed far more than merely monitoring the project; they have contributed many of the substantive ideas incorporated in the research.

Thanks must also go to Jim Mahoney, now with the Department of Health, Education, and Welfare, who during his NASA days was instrumental in connecting us with potential projects for study.

The quality of the manuscript has been improved considerably by the editing of Kathy Piepmeier and John A. Wright, who translated my often cumbersome syntax into more understandable English. Virginia Stupak and Nancy Emery patiently typed the many necessary revisions.

Finally there has been the major contribution of my wife Joan and our children Tommy, Susan Marie, and Máirín. They have not only been very tolerant over the years of work but have contributed in a major way by providing a psychological environment conducive to creative research.

MANAGING THE FLOW OF TECHNOLOGY

1 INTRODUCTION

For the several employments and offices of our fellows, we have twelve that sail into foreign countries under the names of other nations (for our own conceal), who bring us the books and abstracts, and patterns of experiments of all other parts. These we call Merchants of Light.

We have three that collect the experiments which are in all books. These we call Depredators.

We have three that collect the experiments of all mechanical arts, and also of liberal sciences, and also of practices which are not brought into arts. These we call Mystery-men.

We have three that try new experiments, such as themselves think good. These we call Pioneers or Miners.

We have three that draw the experiments of the former four into titles and tables, to give the better light for the drawing of observations and axioms cut of them. These we call Compilers.

We have three that bend themselves, looking into the experiments of their fellows, and cast about how to draw out of them things of use and practice for man's life and knowledge, as well for works as for plain demonstration of causes, means of natural divinations and the easy and clear discovery of the virtues and parts of bodies. These we call Dowry-men or Benefactors.

Then after divers meetings and consults of our whole number to consider of the former labours and collections, we have three that take care out of them to direct new experiments, of a higher light, more penetrating into Nature than the former. These we call Lamps.

We have three others that do execute the experiments so directed and report them. These we call Inoculators.

Lastly, we have three that raise the former discoveries by experiments into greater observations, axioms, and aphorisms. These we call Interpretators of Nature.

We have also, as you must think, novices and apprentices, that the succession of the former employed men do not fail; beside a great number of servants and attendants, men and women. And this we do also: We have consultations, which of the inventions and experiences which we have discovered shall be published, and which not: and take all an oath of secrecy for the concealing of those which we think fit to keep secret: though, some of those we do reveal sometimes to the State, and some not.[1]

Merchants of Light, Depredators, Mystery-men, Miners, Compilers, Benefactors, Lamps, Inoculators, and Interpretators of Nature—they all share a common concern and responsibility for processing information. Information processing is the essence of scientific activity. As physical systems consume and transform *energy,* so too does the system of science consume, transform, produce, and exchange *information.* Scientists talk to one another, they read each other's papers, and most important, they publish scientific papers, their principal tangible product. Both the input and output of this system we call science are in the form of information. Each of the components, whether individual investigations or projects, consume and produce information. Furthermore, whether written or oral, this information is always in the form of human language. Scientific information is, or can be, nearly always encoded in a verbal form.

INFORMATION FLOW IN SCIENCE AND TECHNOLOGY

Technology is also an ardent consumer of information. The engineer must first have information in order to understand and formulate the problem confronting him. Then he must have additional information from either external sources or memory in order to develop possible solutions to his problem. Just like his counterpart in science, the technologist requires verbal information in order to perform his work. At this level, there is a very strong similarity between the information input requirements of both scientists and technologists.

It is only when we turn to the nature of the outputs of scientific and technological activity that really striking differences appear. These, as will be seen, imply very real and important second-order differences in the nature of the information input requirements.

Technology consumes information, transforms it, and produces a product in a form that can still be regarded as information bearing. The information, however, is no longer in a verbal form. Whereas science both consumes and produces information in the form of human language, engineers transform information from this verbal format to a physically encoded form. They produce physical hardware in the form of products or processes.

The scientist's principal goal is a published paper. The technologist's goal is to produce some physical change in the world. This difference in orientation, and the subsequent difference in the nature of the products of the two, has profound implications for those concerned with supplying information to either of the two activities.

The information-processing system of science has an inherent compatibility between input and output. Both are in verbal form (figure 1.1). The output of one stage, therefore, is in the form in which it will be required for the next stage. The problem of supplying information to the scientist thus becomes one of systematically collecting and organizing these outputs and making them accessible to other scientists to employ in their work.

In technology, on the other hand, there is a fundamental and inherent incompatibility between input and output. Because outputs are in form basically different from inputs, they usually cannot serve directly as inputs to the next stage. The physically encoded format of the output makes it very difficult to retrieve the information necessary for further developments. That is not to say that this is impossible: technologists frequently analyze a competitor's product in order to retrieve information; competing nations often attempt to capture one another's weapon systems in order to analyze them for their information content. This is a difficult and uncertain process, however. It would be much simpler if the information were directly available in verbal form. As a consequence, attempts are made to decode or understand physically encoded information only when one party to the exchange is unavailable or unwilling to cooperate. Then an attempt is made to understand how the problems were approached by analyzing the physical product. In cases where the technologists responsible for the product are available and cooperative, this strategy is seldom used. It is much more effective to communicate with them directly, thereby obtaining the necessary information in a verbal form.

Figure 1.1 Information Processing in Science and Technology

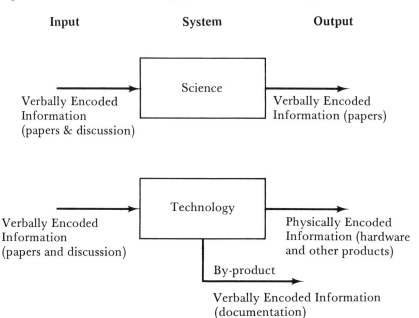

A question that arises concerning the documentation produced in the course of most technological projects is why it cannot serve to meet the information needs of subsequent stages in technological development. The answer is that it is not quite compatible with other input requirements although it meets the requirements of verbal structure. First, as seen in figure 1.1, it is merely a by-product. The direct output is still physical, consequently it is incomplete. It generally assumes a considerable knowledge of what went into the physical product. Those unacquainted with the actual development therefore require some human intervention to supplement and interpret the information contained in this documentation. Thus, technological documentation is often most useful only when the author is directly available to explain and supplement its content.

Now if all of this is true, it leads to an interesting conclusion: whereas the provision of information in science involves the gathering, organizing, and distribution of publications, the situation in technology is very different. The technologist must obtain his information either through the very difficult task of decoding and translating physically encoded information or by relying upon direct personal contact and communication with other technologists. His reliance upon the written word will be much less than that of the scientist. Thus, there are very different solutions to the problems of improving the dissemination and availability of information in the two domains. If, for example, one were to develop an optimum system for communication in science, there is no reason to suspect that it would be at all appropriate for technology. It is essential that we bear these distinctions in mind while exploring the nature of the communication processes in technology. Much has been written about scientific information flow; we may even understand something about it. One must be extremely cautious, however, in extrapolating or attempting to apply this understanding of science to the situation in technology.

THE INFORMATION EXPLOSION

Studies (for example, Price, 1961) showing an exponential growth
rate of scientific journal articles and manpower are too well
known to require explication here, but they provide the stimulus
for much of the recent concern over the enormous amount of
scientific information being generated. The now-famous report
of the President's Science Advisory Committee (1963) noting
this trend warned of the impending danger that science may
fragment into a "mass of repetitious findings, or worse, into conflicting specialties that are not recognized as being mutually inconsistent."

Because technological activity has also increased over the years,
the problem of rapidly expanding knowledge exists in technology
as well. In fact, the problem may even be more serious in technology, which shares the problem of the information explosion
in two ways. First, it is at least partially dependent upon inputs
of scientific information to maintain its progress. Technologists
must therefore find ways to keep abreast of relevant science. In
addition, it has also grown enormously over the years. It has
become extremely difficult to keep up with the state of the art
in many rapidly burgeoning fields of technology. It is for this
reason that the problem of "technological obsolescence" has
thrust itself upon us.

Unlike many physical assets, technological information is not a
dormant body of material, that can be drawn upon as needed.
It is, instead, a dynamic quantity, integral with and absolutely
essential to the process by which it is generated. Research and
development is a process that feeds upon itself, and the communication of its products is of paramount importance to the
proper nurturing of the process. Continued R&D is impossible
without some communication of results. The tremendous increases in recent years in the amount of R&D performed in the
world has resulted in a concomitant increase in the amount of
information to be communicated, presenting the user with the
difficult problem of plowing through a morass of available

information to reach the information pertinent to his problem.

For the most part the response to these difficulties has taken the form of research aimed at improving existing information systems or developing new ones. This end is being accomplished through studies of both hardware and software techniques to improve the means by which information is made accessible to the user. The approach has generally suffered from two faults. First, it has assumed that science and technology are similar and that the solutions developed for science can be applied to technology. A second shortcoming lies in the fact that it is the information suppliers who instigated organized efforts to deal with the problem. These generally have some investment in one or another class of solution and are not interested in solutions that might threaten their established positions. Their efforts are therefore directed toward improving and marketing their own approaches. This presents a classic case of a solution searching for a problem, a strategy almost certainly doomed to failure. Saul Herner (1959) stated this failure to consider the user and his problem:

Perhaps the most important and least considered factor in the design of information storage and retrieval systems is the user of such systems. Regardless of what other parameters are considered in the development of a storage and retrieval mechanism, it is necessary to consider its potential use and mode of use by the persons or groups for whom it is intended; it is necessary either to fashion the system to suit the user's needs, habits, and preferences or to fashion the user to meet the needs, habits, and preferences of the system. Both approaches are possible, but the second one, involving education and re-education of the user, is evolutionary and futuristic. A system designed for now should at least be able to serve the present user.

USER STUDIES

In recognition of this, there have been studies to develop a better understanding of information users and their needs and to determine empirically the manner in which existing communication systems function.[2] These studies of scientific communication in process, known as "user studies," have three general goals (Menzel, 1962):

1. To distinguish the types of informational needs that scientists have and to determine in what respects they remain unsatisfied.
2. To examine the means and occasions of scientific information exchange in order to single out the features that make them more or less able to meet the scientists' several needs.
3. To analyze characteristics of the scientist's specialty, his institution, and his outlook as possible conditions that may influence his needs for information, his opportunities for satisfying them, and, hence, his information-gathering habits and satisfactions.

In sum, the purpose of user studies is to define the problem and thereby provide direction not only to those who are designing improved systems and techniques of communication but to those policymakers whose task is to determine the emphasis and direction that such improvement programs should take. The difficulty has been that while many such user studies have been performed and have generated a correspondingly large volume of data, they have been for the most part piecemeal in their approach. The approach has been fragmentary, and there has been little or no attempt to integrate their results into a more complete systemic understanding of the communication process in science or technology.

THE PRESENT RESEARCH AND PURPOSE OF THIS VOLUME
The present work is an attempt to fill this void. It is an extensive ten-year project, which began in one sense as a user study but which expanded in scope gradually as knowledge of the total system developed. It is a system-level approach to the problem of communication in technology, which looks at not just average users of information but extraordinary users as well. In fact the existence of marked differences in consumption patterns is one of the key findings. The discovery of a wide variance in the manner by which technologists acquire information is what led eventually to the concept of technological gatekeepers, who link their organizations to the technological world at large. Similarly,

a view of the communication system at the system level has led to the consideration of organizational structure and architectural design as important considerations in supplying technological information to the user. These are unanticipated solutions, which would never have occurred if the problem had been approached solely from the supplier side or in a merely piecemeal fashion from the user side. Thus, research reported in this volume can be considered a "user study." In fact, the study goes far beyond Menzel's three goals. The data presented will provide tools that management of R&D organizations can use to keep their staff more current with the state of the art in their technologies.

Because of the nature of technology, there will be far less emphasis on the kinds of storage and retrieval systems that Herner had in mind. Rather, the principal emphasis will be on restructuring of human and organizational systems to bring about better person-to-person contact. The reason for this emphasis will become clear as the volume progresses.

There will be two general themes developed in the course of this book. Chapter 2 describes the research program from which the data are derived, and chapter 3 gives an overview of the communication system in technology. Chapters 4 and 6 will be concerned with the acquisition of technical information by the R&D organization. Chapter 5 first demonstrates and discusses the importance of promoting communication within the organization and then goes on to discuss mechanisms that can be used to improve the dissemination of technical information within the organization. The theme of dissemination is picked up again in chapters 7, 8, and 9, which consider the impact of organizational structure, building architecture, and office layout on communication.

A straightforward sequencing from the problems of information acquisition to those of dissemination is spoiled by having the content of chapter 5 precede that of chapter 6. The nature of the data, however, requires this positioning. One must see the importance of communication within the organization in order to understand completely the most effective means by which the organization can acquire information.

NOTES

1. Sir Francis Bacon (in describing his proposed research laboratory "Salomon's House") in *The New Atlantis* (Johnson, 1965).

2. For a good overview of this very large body of research, see the *Annual Reviews of Information Science and Technology,* edited by Carlos Cuadra, and published by the Encyclopedia Brittanica, especially the chapters on information needs and uses.

2 THE RESEARCH PROGRAM

Research reported in this volume was conducted over a period of about ten years, beginning in 1963 and concluding about 1973. This program of research can be roughly divided into two phases, with a shift of orientation between the phases.

In the first phase, the chief concern was to determine the information consumption patterns of research and development projects so that their needs might be better met. Naturally, the focus during the first phase was the R&D project itself. A number of projects of varying size and duration were studied. Their information consumption patterns were measured, and those in turn were related to the quality of the work performed. The second phase of the research was stimulated by the results of the first phase. In the second phase, the focus shifted from the R&D project to the overall laboratory organization in which the project was performed. The principal goal in the second phase was to determine how information enters and flows through an R&D organization.

The research methods used over the course of the program were many and diverse. There was little in the way of standard technique that could be employed to achieve the desired goals. Methods, therefore, had to be developed on an ad-hoc basis. For this reason, it is proper that a brief description be supplied at this point, so that the reader may be better able to understand and evaluate the results.

PHASE 1: THE STUDY OF PARALLEL OR TWIN PROJECTS

Performance Measurement and the Generalization of Results
Performance measurement is an essential ingredient of any study that purports to help managers learn how to improve the functioning of their organizations. In order to draw normative as well as descriptive conclusions from a study, some measure of performance is obviously necessary. This is certainly not a startling revelation; yet the study that incorporates even a modestly hard measure of performance is still the exception rather than the rule.

The descriptive study measures or describes the behavior of managers or other workers under certain circumstances and merely reports how these people behave in certain types of organizations or circumstances. The results of such studies are often interesting and sometimes valuable, but it is rare that management policy, which must improve organizational performances, can be based on such research. Policy must be based on normative results. Organizational research must, therefore, relate observed behaviors to performance in order to develop results that will be of real value to policymakers. Such research must go beyond the merely descriptive and develop normative statements, which must prescribe the direction for improvement in organizational performance. In order to do this, of course, one must first be able to measure improvement. The reason for this avoidance or evasion of performance measurement lies not in the contrary nature of organizational researchers but in the difficulty of measuring organizational performance.

Although the problem of performance measurement is a general one in organizational research, nowhere is it more salient and difficult than in studies of R&D organizations or of the process of science or technology. In science and technology, each piece of work is, by definition, unique. If the problem has been solved before, it is no longer research. There is no replication of activity in R&D as there is in so many other activities. This absence of replication makes performance more difficult to measure because there is no common denominator among projects to provide a basis for performance comparison. This has led investigators to the use of very detailed case studies in the hope of learning more about the process of scientific or technological research. But the case study has to be merely descriptive in nature. How can one measure performance in a single instance? Performance compared to what? Unfortunately there are no absolute scales, so the case study does not solve the problem of performance measurement; worse still, it introduces an additional problem: generalizing results. Because each case is unique, it is impossible to determine how much of what one learns from the case is also unique and how much could be applied in different circumstances.

One approach to surmounting such a difficulty is the resort to very large sample sizes. In this way, with proper safeguards on sample structure, generality can be assured. This approach, however, can be a very expensive one if any great detail is desired in the measurements to be taken. It usually implies a combination of survey technique and one-point-in-time measurement. Generality is thus purchased at the cost of depth and on-line, real-time measurement of behavior.

The Twin Project Approach

To surmount these intrinsic difficulties, a middle ground was sought, one that would allow much of the depth, detail, and real-time measurement obtainable through case studies while allowing the results to be generalized beyond the instances at hand and enabling a measure of performance to be made. This was done by locating instances in which there was a limited replication of R&D projects. These were cases in which there was a duplication of effort between two or more R&D teams. Two or more groups or projects were established to solve the same set of technical problems.

This matched case approach to studying technical problem solving by engineers and scientists is similar to the research strategy usualy employed by experimental psychologists in the study of human problem-solving behavior. By assigning the same problem to large numbers of subjects, the psychologists hold the substance of the problem constant and can make a valid comparison of the subjects' approaches. This approach is also similar to the way in which environmental and genetic influences are separated in the study of human personality development through the use of identical twins exposed to different environmental influences.[1] This technique of matched pairs has also been used to great effect in agricultural research. Beveridge (1957), for example, reports that studies of butterfat yield in New Zealand cattle show that "as much information was obtained per pair of identical twin cows, as from two groups each of 55 cows!"[2]

After describing how unique R&D efforts are. it may seem inconsistent to say that such instances of replication are not rare. There are, in fact, many other instances in which more than one R&D group takes on the same set of problems at the same time. This situation often results from competitive pressures to produce a product to fulfill some market need that has been perceived simultaneously by a number of firms. Another source of such replication stems from the fact that in recent years government agencies and corporations have recognized that a broader view of a problem and its potential solutions can often be gained through the conduct of parallel efforts. Consequently, government agencies often contract with two or more firms to provide preliminary design studies or even prototype hardware to meet certain prescribed needs. The replication that this latter situation produces allowed us to study two or more teams working on the same set of problems and then to have their solutions evaluated by the technical personnel in the government laboratory, whose responsibility it was to monitor and evaluate the work.

The matched twin approach enabled us to simultaneously solve the problems of generality of results and performance measurement as well. Performance in R&D is difficult to measure primarily because of the degree to which each problem is unique. One cannot validly compare solutions to different problems in terms of their quality. One can make much more valid comparison, however, of solutions that are intended for the same problem. The comparative evaluation of the relative quality of different solutions to the same problem can thus be assessed with much greater confidence and validity than any evaluation of single solution to a problem determined on an "absolute" basis. In the case of R&D problems, this is all the more true when prototype hardware is produced and evaluations can be based on physical tests of the prototypes. This was the actual case in some of the instances that we studied. In the remaining cases, technical approaches were proposed and evaluated by either simulation or judgment on the part of the government's technical monitors. The relative judgment approach is, of course, the least reliable. In a very large proportion of these, however, the evaluators were sufficiently candid to tell us that they were unable to make a distinction in quality. This meant that in many cases no performance measure was available, a frustrating situation. But it also meant that our confidence could be that much higher in those cases where the evaluators saw a distinct difference in solution quality. They were not arbitrarily evaluating to keep some social scientists happy; they were instead making an honest effort to meet our needs with rigorous, informed evaluations of the work.

Forms of Data Collected During the Study of Projects
In order to determine the information consumption patterns of the projects, data had to be collected in a variety of ways. Before they could be studied, however, projects had to be located, and their substance had to be understood. Instances in which parallel or twin contracts were to be awarded were located through *Commerce Business Daily* and by notification by certain government agencies that were interested in our study. Once a planned project award was found, an effort was made to determine as much as possible about the nature and substance of the work to be performed. This work was done by analysis of the contract work statement and discussion with technical personnel in the agency awarding the contracts. Because the government had to describe to the contractors what it was that they were to do under the contract, they had to draw up a fairly detailed statement of work. This statement was obtained and then analyzed and factored into a reasonable number of subproblem areas (generally subsystems). The resulting breakdown was checked with the technical person who prepared the work statement, and corrections were made. Forms for the collection of data from the project teams were then designed. Data were gathered by four means:

1. Time allocation forms completed by individuals provided information on a daily basis regarding the total amount of time spent working on a given subproblem, as well as the time spent in a number of information-related activities.
2. Forms that became known as solution development records provided valuable indexes of problem solvers' progress toward solutions. This particular instrument will be described in more detail later in the chapter.
3. Detailed interviews were held with individual problem solvers before and after the project, and telephone interviews were made at points during the project when it appeared that some critical change was occurring.
4. Periodic tape-recorded progress reports by some of the project managers provided useful general background information.

Time Allocation Forms Used on a few of the projects, these were completed at the end of each day and showed both the total amount of time spent working on a specified technical problem and the breakdown of that time into the following broad categories:

1. Literature search.
2. Consultation with colleagues within the organization.
3. Consultation (formal or informal) with persons outside the organization.
4. Analytic design.

An example of this form is shown in the appendix to this book.

Solution Development Records One of the most critical and difficult problems faced in doing field research of the type reported here is that of gaining the full cooperation of the participants. The researcher must be able to assure participants that their cooperation is matched by the benefits they will derive from this cooperation. Because the only benefits that could be offered during the project study were in such long-range terms as the general contribution to knowledge or potential future

changes in the effectiveness of managing R&D organization, it
was incumbent upon us to reduce the cost of participation to
an absolute minimum. We did this by developing the solution
development record, a simple device to monitor the progress of
problem-solving activities and to relate this progress to the receipt
of technical information. This was a tall order. Fortunately,
however, there appeared to be a critical concept—which seemed
amenable to simple operational measurement—that linked together both problem activity and information receipt.

The concept of subjective probability, which can be defined as
a relationship that exists between hypothesis and evidence (Carnap, 1962), became the link between problem solving and communication. In a technological problem, if one were to consider
solution alternatives as hypotheses and have the problem solver
ascribe (subjective) probabilities to these, then any change in
probability could be attributed to a change in the state of evidence,
or knowledge, on the part of the problem solver. Such a change in
knowledge is a change in information state and must have occurred
through the introduction of new information. This introduction
might have come from memory, when a piece of information
already held in memory enters the consciousness for the first
time. The information might also have come from outside the
problem solver from a variety of sources. It is the receipt of
such external messages and the determination of their sources
that we set out to discover with the solution development record.

The solution development record has developed into a research
tool that provides a record over time of the progress of an indi-
dual engineer or group of engineers (or scientists) toward the
solution of a technical problem. The individual engineer, or
the lead engineer in the case of a group,[3] responsible for each
subproblem is asked to provide a weekly estimate of the probability that each alternative approach under consideration will be
finally chosen as the solution to that particular subproblem
(figure 2.1).

If at some point in the design the person working on the problem

Figure 2.1 A Fictitious Example of a Solution Development Record

Solution Development Record
Manned Uranus Landing in an Early Time Period Study

Name _____ Date _____

Subproblem no. 1: Design of the
electrical power supply subsystem
for the space vehicle   Estimate of Probability that Alternative
                                              will be Employed
a. Alternative approaches:
    hydrogen-oxygen fuel cell 0 0.1 0.2 0.3 0.4 0.5 0.6 0.7 0.8 0.9 1.0
    KOH fuel cell   0 0.1 0.2 0.3 0.4 0.5 0.6 0.7 0.8 0.9 1.0
    Rankine cycle thermal
      reactor   0 0.1 0.2 0.3 0.4 0.5 0.6 0.7 0.8 0.9 1.0
    Brayton cycle reactor   0 0.1 0.2 0.3 0.4 0.5 0.6 0.7 0.8 0.9 1.0
    _____ 0 0.1 0.2 0.3 0.4 0.5 0.6 0.7 0.8 0.9 1.0
    _____ 0 0.1 0.2 0.3 0.4 0.5 0.6 0.7 0.8 0.9 1.0

b. If information which had impact upon your visualization of the problem
    or any of its potential solutions was received at any time during the past
    week, please circle the source(s) of that information on the line below.
    Sources defined on the reverse side.
    Information Source: L V C ES TS CR PE E
    comments (if any): _____

in figure 2.1 were considering two technical approaches for providing electrical power to the space vehicle and he were completely uncommitted between the two, he would circle 0.5 for each.[4] Eventually, as the solution progresses, one alternative will attain a 1.0 probability, and the others will become zero. The probabilities plotted over time become a graphic record of the solution history.

Alternative approaches are identified from the contract work statement when so specified or from the responsible engineer when he is interviewed prior to beginning the task. Blank spaces are always provided so that new approaches may be reported as they arise. In cases where a respondent believed there was some probability of choosing an approach that he could not specify at the time, he was instructed to assign a probability to an approach labeled "other."

A section at the bottom of the form allowed the respondent to report the source of any important messages that he might have received during the week. These data were not used in any direct measurement of information source usage but were used instead to aid in interpreting probability changes and to prompt the respondent's memory during the postproject interview. The symbols on the figure are decoded in table 2.1 in the same way that they were presented to respondents on the back of the solution development record.

Over the course of a project, a copy of the solution development record was mailed every week to each respondent. The forms are sufficiently flexible so that new alternatives may be incorporated, old ones dropped, and in instances in which an early solution is reached and "frozen," subproblems at the next level could be substituted.

The solution development record, by economizing on the respondent's time, provides an efficient record of a project's history. When major changes in probability occur or when new ideas are introduced, the respondent can be called for a brief telephone interview in order to determine what brought about

Table 2.1 Definitions of Information Sources Presented to the Problem Solvers

L = literature	Books, professional, technical, and trade journals, and other publicly accessible written material.
V = vendors	Representative of, or documentation generated by, suppliers or potential suppliers of design components.
C = customer	Representatives of, or documentation generated by, the government agency for which the project is performed.
ES = external sources	Sources outside the laboratory that do not fall into any of the above three categories. These include paid and unpaid consultants and representatives of government agencies other than the customer agency.
TS = technical staff	Engineers and scientists in the laboratory not assigned directly to the project being considered.
CR = company research	Any other project performed previously or simultaneously in the lab regardless of its source of funding.
PE = personal experience	Ideas that were used previously by the engineer for similar problems and are recalled directly from memory.
E = experimentation	Ideas that are the result of tests or experiments with no immediate input of information from any other source.

the change. When the project is completed, each respondent is presented with a time plot of his probability estimates and is interviewed at some length to further determine causes and effects of the design changes reflected in this record. The solution development record thus is a mechanism for monitoring the project in a relatively unobtrusive manner and provides, through its time plot, a stimulus to the individual's memory once the project is ended. This latter function is extremely important because more time was available for detailed interviews after the project had ended. The plot aided the engineer in recalling project events and activities, and it probably also prevented some ex-post rationalization from occurring. The hard record of reports could not be ignored, and peculiarities in the data had to be explained.

Data Obtained from Interviews Cooperating engineers and scientists were interviewed prior to the initiation of a project in order to determine whatever alternatives they were considering for solutions to their problems and to obtain a feeling for the manner in which a person's previous experience might relate to the project at hand.

During the project some engineers were telephoned to determine the causes of major probability shifts or the origins of new ideas that had been introduced. In addition to explaining the specific items, these interviews provided information that was extremely valuable in understanding subsequent events.

Following completion of a project, each respondent was presented with the time plot of his solution development record reports and was interviewed in depth to determine the causes of each change in the probability level assigned to an alternative. Also sought were the reasons for discarding alternatives and the information sources that were used throughout the study for such functions as the generation of alternatives, the further determination of the nature and definition of the problem, and so on. The length of the interview generally ranged from a half-hour to an hour per person, but a few ran on for several hours.

The interviews, including telephone interviews, were all tape recorded and were later coded to determine, among other things, the information sources used for each of several problem-solving functions.

A Typical Problem To further illustrate the research method and the nature of the data obtained, a single subproblem has been selected from one of the projects studied, and the solution process for the subproblem will be described in some detail. The time plot provided by the solution development record provides a rather interesting perspective on the history of this solution process (figure 2.2) and illustrates the intimate relation between the flow of technical information and technical problem solving.

The figure shows the approaches two engineers followed in two different organizations, Laboratory Y and Laboratory Z, in the design of the reflector surface for a very large antenna. While

Figure 2.2 A Plot of Solution Points Over Time for a Problem Involving the Design of the Reflector Portion of a Large Antenna System

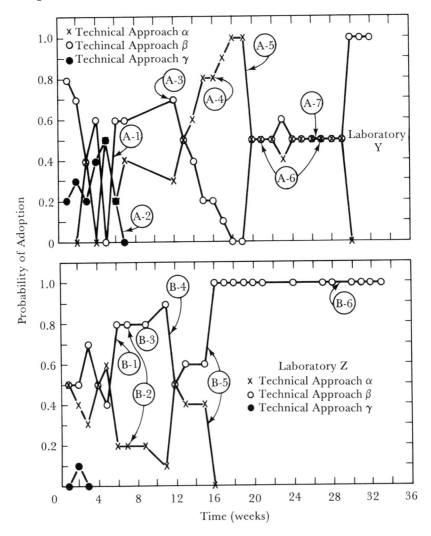

both engineers ultimately decided upon the same general approach, they arrived there by quite different routes. A brief look at the kinds of information obtained by the two problem solvers (indicated in the figure by flagnotes) will explain much of this difference.

Early in the study, as indicated by flagnotes A-1 and B-1, the customer (a government laboratory) supplied both contractors with the results of an experiment to determine the loadings that the antenna would experience at different wind velocities. Judging from these data, approach β would appear to create reasonably acceptable wind loadings, and it consequently rose in favor at both laboratories. Prior to this time, Laboratory Y showed considerable vacillation among the three alternatives; Laboratory Z, during this early period, conducted an intensive literature search but failed to uncover any evidence of empirical or analytical work dealing with aerodynamic shapes anything like that to be used in the antenna, particularly at the low air speeds under consideration.

At A-2 Laboratory Y's aerodynamic staff performed an analysis that showed wind load moments for approach γ to be about twice that for α or β. At the same time, the electromagnetic staff reported that while approaches α and β satisfactorily met electrical requirements up to 3 gigahertz, β provided better amplification at the higher frequencies.

A similar electromagnetic analysis performed by Laboratory Z shows that approach α fails to perform as well as β in terms of antenna gain. This is indicated by flagnote B-2. During the same period (B-3), this laboratory attempted to extrapolate from previously acquired data (an earlier antenna) to estimate wind loads.

About the twelfth week (A-3) Laboratory Y conducted a wind tunnel study, which showed that approach β resulted in a wind torque considerably larger than that predicted by the customer's data (A-1).[5] Since approach α did not perform as well as β electrically, one of three things now had to be done: the alternative providing the second-best amplification characteristics had to be accepted, the aerodynamic specification had to be relaxed, or a new alternative, which would meet both the aerodynamic specification and the amplification of approach β, had to be generated. Because a new alternative did not present itself, negotiations were begun in order to determine the customer's position on the first two solutions.

The customer decided to accept the reduced amplification (A-4), and approach α rose in favor accordingly. The change in the customer's requirements was provided to Laboratory Z as well, but there is no indication that this information changed the way in which the staff viewed approach β or the nature of the problem itself.

Laboratory Z's failure to respond undoubtedly resulted from the fact that Z did not have as complete information as Y regarding approach β. The brief drop in β's position at Laboratory Z (B-4) was a result of some doubts this laboratory had concerning the feasibility of the approach, but as far as can be determined the doubts were not based upon any hard data. Following this brief period of skepticism β rose rapidly to a 1.0 level (B-5) and was further established there when, as indicated by B-6, information concerning special fabricating machinery became available.

At Laboratory Y, meanwhile, approach α encountered some difficulty with the cost analysts, resulting in the drop in probability (indicated at note A-5). Laboratory Y remained indifferent between approaches α and β for quite some time while trade-off studies were pursued (A-6). Numerous contacts were made with vendors during this period to determine the costs associated with different elaborations of the two approaches.

Finally (A-7) information was obtained from the National Weather Service, which showed that a relatively minor change in the location of the antenna installation would considerably reduce the antenna's exposure to high velocity winds. The new site allowed a 20 percent reduction in the wind-loading specification, while retaining the requirement that the antenna be out of operation, due to weather, a maximum of 5 percent of the time. This information was instrumental in Laboratory Y's decision to adopt approach β.

The work statement for the subsystem under consideration suggested three technical approaches, and these were the only ones the two teams considered. This is not the usual case. Competing engineers seldom consider the same approaches to a problem, and often they consider far more than three. A more typical case is illustrated in figure 2.3. In this instance, each engineer considered five alternative approaches to the problem. Three of these (δ, ϵ, and ζ) were specified by the customer at the start of the contract. Two more were generated by each team during the course of the project. In Laboratory Y, approach η had been previously used by the engineer on problems of a similar nature, and the association of this problem with the former ones brought the idea out of memory.

Approach θ, which was finally chosen as the solution, has an interesting origin. Obviously from the figure, the problem solver was not satisfied with any of the possibilities he had considered until the time that θ first appeared. He apparently made this dissatisfaction known to a colleague who subsequently attended an engineering society conference. At the conference he learned of a new technique, which had been recently developed, that might be of some relevance to the problem solver. When he returned from the conference, he described the technique as well as he could and provided the name of the man who had developed it, as well as some literature references that he had also obtained. The engineer searched out the literature references and contacted the originator of the technique as well. From him he learned more of its potential and limitations and the nature of the adaptations he would have to make to put it to use. He then contacted several firms that could fabricate the necessary hardware, got more information from them, and eventually produced a workable physical realization of the technique.

Similar, if less complicated, situations occurred in Laboratory Z, where an engineer, thumbing through a colleague's reference file of clippings from trade journals, ran across an item that suggested approach κ. This representative happened to visit a person in

Figure 2.3 A Plot of Solution Points Over Time for a Problem Involving the
Design of the Scanning Controls for a Large Antenna System

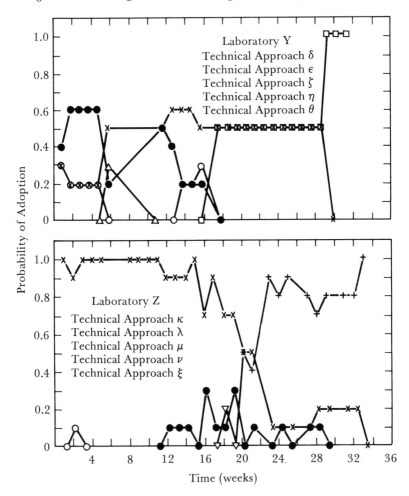

another department who knew of the engineer's problem and suggested that the vendor contact him. The vendor representative did and provided him with additional information, which enabled him to more fully develop approach κ. On another occasion, it was fortuitous contact with a vendor that produced approach λ. The vendor representative happened to visit a friend of the problem solver in another department. The friend knew of the first man's problem and suggested that the vendor contact him.

Units of Analysis Used in the Study of Projects

Several approaches will be taken to determine the information requirements of R&D projects. Data from the twin projects will be analyzed using both man-hours devoted to use of an information channel and messages received via each information channel as the principal units of analysis.[6]

Since engineering man-hours expended is quite closely correlated with cost on projects such as these, this choice of unit provides a measure not only of the relative extent to which channels are used but also of a major portion of the relative expense involved in such use.

Idea-Generating Messages Received

Message refers to a unit of information, either transmitted or received. Menzel (1960) in his methodological review describes this measure in the following way:

> . . . [another] way of delineating the units of observation or recording uses as the basic unit a "message"—i.e., some kind of unit of information transmitted or of communication achieved. Instead of focusing on some piece of scientist's behavior and then, perhaps, asking what information it yielded, it singles out pieces of information and asks whether and through what behavior they were obtained. The neatest embodiment of this approach would be actual studies of the diffusion of a message.

In the present case, the suggestion of a potential solution to the problem or the suggestion of a reason for discarding a solution is considered a message received. A message is not necessarily a complete idea. Usually a potential solution results from messages received through several channels. The engineer integrates these into what becomes a new technical idea. This is where the element of creativity enters. The engineer is not merely regurgitating once-consumed information. He associates messages with one another, sees possible new combinations, and thereby creates something that is new and potentially useful. This process is exemplified clearly in the problems described in the last section. In the first case, for instance, the engineer received messages from a colleague, from the literature references, from the originator of the technique, and from the vendors he contacted. The engineer integrated the information contained in the different messages and developed the final idea from them. This is the normal way in which new ideas were encountered. They seldom came full-blown from a single source. It was more common for several sources to suggest parts of the idea and for the engineer to perform the necessary integration. The engineer's ability to perform this integration is what we have come to call creativity. He performs the creative act, but he is heavily dependent upon information inputs in doing so.

In the interviews, each idea or alternative that the engineer considered was traced back to its sources. This was done while listening to the tape-recorded transcriptions[7] and coding them for messages received from specific channels. When several channels contributed to a single potential solution, credit was given to each source.

The Designation of Specific Communication Channels

In analyzing the data from the time allocation forms, three broad classes of information channel are considered and a measurement is made of time spent with literature, time spent in personal contact outside the laboratory, and personal contact within the laboratory. In the case of "messages received," a finer division is possible, and the interview tapes were coded to determine which of six possible channels was responsible for each message (table 2.2).

Characteristics of the Sample

There are several sample limitations that must be kept in mind while considering the data to be presented in subsequent chapters.

Paramount among these are the size of the laboratories in which
the R&D project teams were embedded, the nature of the principal industries in which the laboratories are themselves situated,
and the general nature of the work performed.

The outstanding characteristic of the laboratories represented
in the study is their size. They were all very large. The smallest
had nearly a thousand technical professional employees; most had
several thousand. This is probably not a serious limitation. In fact,
it probably makes the results conservative in many ways. Many
of the problems observed through the data are probably more
serious in smaller firms. The problem of communicating outside
the organization is one example. Small firms are often forced,
by virtue of their size, to become heavily dependent upon external
sources of information. On the other hand, smaller firms do not
have the same problems of structuring communication networks
that are encountered in larger organizations. The reader who is
concerned primarily with smaller firms must take these facts into
account in weighing the results. Fortunately, it will usually be
obvious from the presentation, when the problem is one of little
or no concern to the smaller organization.

Because all of the projects studied were financed by the United
States government, the first phase of the study is necessarily biased
toward firms whose business interests are slanted toward this type
of customer. In a word, the sample of projects was drawn mostly
from the aerospace and electronics industries. The sample would
be broader if we had been able to include commercial chemical,
data-processing equipment, and pharmaceutical projects. However,
some computer and chemical firms were represented, and since
the data show little interindustry difference, we firmly believe
that most of the results can be generalized with little difficulty.

The majority of the projects did not involve direct hardware
developments. They were for the most part studies: either preliminary design studies, feasibility studies, or the study of some
phenomenon having possible long-range potential. Never was their
final product more than a prototype piece of hardware. They were

Table 2.2 Information Channels Considered in the Study

Literature	Books, professional, technical, and trade journals, government publications, unpublished reports, and other written material.
Vendors	Representatives of suppliers of design components or subsystems.
Customer	Representatives of, or documentation generated by, the government agency for which the project is performed.
External sources	Sources outside the laboratory that do not fall into any of the above three categories. These include paid and unpaid consultants, representatives of government agencies other than the customer agency, university faculty, or representatives of firms other than vendor firms.
Technical staff	Engineers and scientists in the laboratory not assigned directly to the project being considered.
Company research	Any other project performed previously or simultaneously in the laboratory.

never intended to lead *directly* to production efforts but were several phases earlier and were sometimes, but not always, followed up by further study or development contracts awarded to one or more of the competing laboratories. The projects, in other words, represent a very early conceptual phase in the technological process and were to a great degree devoid of the real-life nuts-and-bolts type of hardware problem. This is not to say that they were not hardware oriented at all. All but two, which were feasibility studies, were intended to result eventually in some hardware development. Hardware constraints were therefore accorded their due importance but actual fabrication was generally not involved, and the unique problems that arise only when this point has been reached were noticeably absent. Perhaps the best way to illustrate this is to prôvide brief descriptions of the projects themselves. A total of thirty-three project teams worked on the following seventeen general problems:[8]

1. The design of the reflector portion of a very large and highly complex antenna system for tracking and communication **with**

space vehicles at great distances (two project teams).
2. The design of a vehicle and associated instrumentation to roam the lunar surface and gather descriptive scientific data (two project teams).
3. An investigation of possible mission profiles and propulsion techniques for manned flights to another planet (two project teams).
4. The development of a detailed mathematical cost model yielding complete R&D and operational program cost estimates of space launch vehicle systems (two project teams).
5. The preliminary design of an earth-orbiting space station (one project team).
6. The design of a deep space probe and appropriate instrumentation (three project teams).
7. The preliminary design of an interplanetary space vehicle (two project teams).
8. The preliminary design of a special-purpose manned spacecraft for cislunar missions (two project teams).
9. The preliminary design of an earth-orbiting space laboratory (two project teams).
10. The development of a mathematical model and simulation of an interplanetary transportation system (one project team).
11. The development of a container for cryogenic fluids (one project team).
12. The development of a low-thrust rocket engine for maneuvering manned spacecraft (two project teams).
13. The design of an unmanned station to gather scientific data from a position on the lunar surface and transmit it back to earth (three project teams).
14. The development of a system to measure physiological responses to space flight (two project teams).
15. A study to determine certain biological requirements of extended manned space flight (two project teams).
16. Design and development of a valve system for bipropellant rocket systems (two project teams).

17. The development of a technique for transferring modulation between coherent light beams without the addition of energy (two project teams).

PHASE 2: STUDIES OF COMMUNICATION NETWORKS STRUCTURE

In its second phase, the research program turned to the study of the person-to-person communication networks in a number of R&D laboratories. The purpose was to understand better the manner in which new technology or technological information entered and was disseminated among the technical staff. In order to do this, a measure had to be made of the structure of a laboratory's communication network. Since it was the direct person-to-person network that was of interest, some measurement had to be devised of the pattern of interpersonal communication. The method eventually developed was to ask people, by means of an interview or a questionnaire, to name those colleagues with whom they most frequently discussed technical or scientific matters.[9] Individuals could then be connected into a network on the basis of whether they communicated regularly. Contact outside the organization was determined in a similar manner by asking questions about both written and person-to-person contact on a regular basis.

This approach was later modified to one in which actual communications were sampled over time. On randomly chosen days, questionnaires were administered that asked people to recall their communications over the course of that particular day. The questionnaires were given out late in the afternoon of the chosen day and contained a listing of the staff members of the organization.[10] Respondents were asked to check off the names of those individuals with whom they had communicated (about scientific or technical matters) over the course of the given day. The sampling was repeated weekly for periods ranging from five weeks to a year. Networks could then be constructed based on some preestablished communication frequency (weekly or monthly for example).

Network studies were conducted in thirteen different laboratories:
1. Laboratory A is the research department of a small high-
   technology firm in materials science (34 professionals).
2. Laboratory B is one of the research departments of a medium-
   sized aerospace firm (48 professionals).
3. Laboratory C is a social science department of a major university (57 professionals).
4. Laboratory D is part of a university medical school (52 professionals).
5. Laboratory E is the advanced technology department of a
   major aerospace firm (400 professionals).
6. Laboratory F is the central research laboratory of a highly
   diversified firm, principally in the chemical industry (46
   professionals).
7. Laboratory G is the research department of a small chemical
   firm (28 professionals).
8. Laboratory H is an agriculture research organization in a
   European country (170 professionals).
9. Laboratory I is a product development department in a major
   computer firm (20 professionals).
10. Laboratory J is a part of a university-affiliated laboratory in
    the aerospace field (40 professionals).
11. Laboratory K is a department of a major electronics firm
    (138 professionals).
12. Laboratory L is a government research center in the aerospace
    field (120 professionals).
13. Laboratory M is a consulting firm in the field of materials
    science in a European country.
A more complete description of the data gathered from these
thirteen laboratories is provided in chapters 6 and 9.

SUMMARY
The two main phases of the research, which are reported in the
remainder of this volume, are the twin project studies and the

laboratory communication network studies. Chapter 2 has attempted to provide a broad overview of the nature of these two components and some idea of the research methods used. There was, however, great variation within the components. Solution development records, for example, were not used on all of the parallel projects, and in the case of the network studies, many independent variables were measured at different times. There are also a number of subsidiary studies that do not fit neatly into either category. All of these will be reported later, and the methods that were used will be described at the appropriate points. We can now move to the actual reporting of the data.

NOTES

1. Usually as the result of separate adoptions of orphaned identical twins.

2. While we have never attempted to compute such a ratio for the study of twin R&D projects, it is hoped that we have been able to gain a similar degree of leverage.

3. When reports were obtained on a group basis, the group seldom comprised more than two engineers.

4. Since there are two reasons why an engineer might give a 50-50 response like this (both alternatives equally satisfactory or both alternatives are equally unsatisfactory and he was unable to develop a third alternative), a section was added to later versions of the solution development record to resolve this dilemma. See the appendix to this book for a more complete version of the instrument.

5. Laboratory Y was an aerospace firm, which at the time was attempting to diversify into the antenna industry. For this reason it had good wind tunnel facilities readily available. The government laboratory was an electronics laboratory, and had no satisfactory wind tunnel available near their facility. They therefore attempted to determine wind loadings by other means, which, as it turned out, introduced some critical inaccuracies.

6. See Menzel (1960) for a discussion of units of analysis that have been and might be employed in studies of information use with a critical examination of each possibility.

7. Intercoder reliabilities ran in the order of 90 percent.

8. Cooperation was obtained from only one of the two teams working on projects 5, 10, and 11.

9. A minimum frequency of at least once a week was prescribed.

10. In very large organizations, groups rather than staff names were listed, with space to write the actual names of the individuals with whom communication occurred.

3 THE COMMUNICATION SYSTEM IN TECHNOLOGY: AN OVERVIEW

Because the remainder of the volume is devoted primarily to the study of engineers, a few clarifying points should be made at the outset about the characteristics of this body of professionals.

DISTINGUISHING ENGINEERS FROM SCIENTISTS

Engineers are not scientists. Few would contest this statement, and yet the failure to recognize the distinction has created untold confusion in the literature. Despite the fact that they should be the last to commit such an egregious error, social scientists studying the behavior of scientists and engineers seldom distinguish properly between the two groups. The social science literature is replete with studies of "scientists," who upon closer examination turn out to be engineers. Worse still, in many studies the populations are mixed, and no attempt is made to distinguish between the two subsets.[1] Many social scientists still view the two groups as essentially the same and feel no need to distinguish between them. This sort of error has led to an unbelievable amount of confusion over the nature of the populations that have been studied and over the applicability of research results to specific real-life situations. The usual practice is to use the term *scientist* throughout a presentation, preceded by a disclaimer to the effect that "for ease of presentation, the term *scientist* will be assumed to include both engineers and scientists." This approach totally neglects the vast differences between the two professions. One might almost as readily lump physicians with fishermen. The practice is especially self-defeating in information use studies because confusion over the characteristics of the sample has led at times to what would appear to be conflicting results and to great difficulty in developing normative measures for improvement of the information systems in either science or technology.

At this point, many readers will accuse the author of magnifying what they may consider a trivial issue. But it is just that failure to recognize the distinction that has resulted in so much misdirected policy. In the field of information science, it has often resulted in heavy investments in solutions to the wrong problem. Engineers

differ from scientists in their professional activity, their attitudes,
their orientations, and even in their typical family background.
To interpret the results of research, it is essential to know whether
those results were derived from the study of engineers or of
scientists because the behavior of the two is so different.

With but a few notable exceptions, studies of information use
have followed this general pattern in not treating engineers separately. Even the chapters devoted to user studies in the *Annual Review
of Information Science and Technology* for the most part fail to
distinguish properly between studies in which the sample population comprised engineers and in which the population consisted
of scientists. The confusion is largely due to the initial emphasis
on scientific information with little explicit regard for the information problem in technology. The "information explosion" was
first recognized by scientists or former scientists, and as a result
engineers were usually included in studies only by accident. They
were normally included in samples only when they represented
a small portion of a more general population of scientists or when
a group of them just happened to be more readily available to an
investigator interested in studying the use of "scientific" information flow. In fact there are still very few studies directed exclusively and explicitly at the communication behavior of engineers.

Engineers and scientists, despite surface similarities, are so fundamentally different in their natures that one could hardly expect
similarity in communication behavior. Not only are the two groups
socialized into entirely different subcultures but their educational
processes are vastly different, and there is a considerable amount of
evidence to show that they differ in personality characteristics and
family backgrounds as well. Krulee and Nadler (1960) contrast
the values and career orientation of science and engineering under-
graduates in the following ways:

[Students] choosing science have additional objectives that distinguish them from those preparing for careers in engineering and
management. The science students place a higher value on

independence and on learning for its own sake, while, by way of contrast, more students in the other curricula are concerned with success and professional preparation. Many students in engineering and management expect their families to be more important than their careers as major sources of satisfactions, but the reverse pattern is more typical for science students. Moreover, there is a sense in which the science students tend to value education as an end in itself, while the others value it as a means to an end.

Note that Krulee and Nadler do not distinguish between engineering students and students in management. There is considerable evidence to show that many engineering students see the profession as a transitional phase in a career leading to higher levels of management. There is evidence also that many of them are successful in accomplishing their long-term career objectives (Schein & Bailyn, 1975). Krulee and Nadler go on to describe the orientations of engineering students:

Engineering students are less concerned than those in science with what one does in a given position and more concerned with the certainty of the rewards to be obtained. It is significant that they place less emphasis on independence, career satisfactions, and the inherent interest their specialty holds for them, and place more value on success, family life, and avoiding a low-level job. On the whole, one suspects that these students want above all for themselves and their families some minimum status and a reasonable degree of economic success. They are prepared to sacrifice some of their independence and opportunities for innovation in order to realize their primary objectives. They are more willing to accept positions which will involve them in complex organizational responsibilities and they assume that success in such positions will depend upon practical knowledge, administrative ability and human relations skills.

In the same vein, Ritti (1971) finds a marked contrast between the work goals of scientists and engineers after graduation (table 3.1). Ritti draws the following three general conclusions from the data of his study:

First, the notion of a basic conflict in goals between management and the professional is misapplied to engineers. If the goals of the

Table 3.1 Work Goals of Research Scientists and Engineers

Work goal: How important is it to you to —	Percentage Indicating Goal is "Very Important"	
	Scientists[a] (N = 33)	Engineers[a] (N = 4,582)
have the opportunity to explore new ideas on technology or systems	b	61
have the opportunity to help the company increase its profit	28	69
gain knowledge of company management policies and practices	19	60
participate in decisions that affect the future business of the company	6	41
work on projects that have a direct impact on the business success of your company	b	47
advance to a policy-making position in management	6	32
work on projects you yourself have originated	75	32
establish your reputation outside the company as an authority in your field	84	29
publish articles in technical journals	88	15
be judged only on the basis of your *technical* contributions	b	13

[a] The relative size of the two samples reflects the size of the two populations in one very large, very technology-intensive firm.

[b] This item is not included in survey of this group (from Ritti, 1971).

business require meeting schedules, developing products that will be successful in the marketplace, and helping the company expand its activities, then the goals of engineering specialists are very much in line with these ends.

Second, engineers do not have the goals of scientists. And evidently they never had the goals of scientists. While publication of results and professional autonomy are clearly valued goals of Ph.D. scientists, they are just as clearly the least valued goals of the baccalaureate engineer. The reasons for this difference can be found in the work functions of engineers as opposed to research scientists. Furthermore, both groups desire career development or advancement, but for the engineer advancement is tied to activities within the company, while for the scientist advancement is dependent upon the reputation established outside the company.

The type of person who is attracted to a career in engineering is fundamentally quite different from the type who pursues a scientific career. On top of all of this lies the most important difference: level of education. Engineers are generally educated to the baccalaureate level; some go on to a Master of Science degree; some have no college degree at all. The scientist is almost always assumed to have a doctorate. The long, complex process of academic socialization that is involved in reaching this stage is bound to result in a person who differs considerably in his lifeview. These differences in values and attitudes toward work will almost certainly be reflected in the behavior of the individuals. To treat both professions as one and then to search for consistencies in behavior and outlook is almost certain to produce error and confusion of results.

THE NATURE OF TECHNOLOGY

The differences between science and technology lie not only in the kinds of people who are attracted to them; they are basic to the nature of the activities themselves. Both science and technology develop in a cumulative manner, with each new advance building on and being a product of vast quantities of work that have gone before. In science all of the work up to any point can be found permanently recorded in literature, which serves as a repository for all scientific knowledge. The cumulative nature of science can be demonstrated quite clearly (Price, 1965a, 1970) by the way in which citations among scientific journal articles cluster and form a regular pattern of development over time.

A journal system had been developed in most technologies that in many ways emulates the system originally developed by scientists; yet the literature published in the majority of these journals lack, as Price (1965b, 1970) has shown, one of the fundamental characteristics of the scientific literature: it does not cumulate or build upon itself as does the scientific literature. Citations to previous papers or patents are fewer and are more often to the author's own work. Publication occupies a position of less importance than it does in science where it serves to document the end product and establish priority. Because published information is at best secondary to the actual utilization of the technical innovation, this archival is not as essential to ensure the technologist that he is properly credited by future generations. The names of Wilbur and Orville Wright are not remembered because they published papers. As pointed out in chapter 1, the technologist's principal legacy to posterity is encoded in physical, not verbal, structure. Consequently, the technologist publishes less and devotes less time to reading than do scientists.

Information is transferred in technology primarily through personal contact. Even in this, however, the technologist differs markedly from the scientist. Scientists working at the frontier of a particular specialty know each other and associate together in what Derek Price has called "invisible colleges." They keep track of one another's work through visits, seminars, and small invitational conferences, supplemented by an informal exchange of written material long before it reaches archival publication. Technologists, on the other hand, keep abreast of their field by close association with co-workers in their own organization. They are limited in forming invisible colleges by the imposition of organizational barriers.

BUREAUCRATIC ORGANIZATION

Unlike scientists, the vast majority of technologists are employed by organizations with a well-defined mission (profit, national defense, space exploration, pollution abatement, and so forth). Mission-oriented organizations necessarily demand of their technologists a degree of identification unknown in most scientific circles. This organizational identification works in two ways to exclude the technologist from informal communication channels outside his organization. First, he is inhibited by the requirements that he work only on problems that are of interest to his employer, and second, he must refrain from early disclosure of the results of his research in order to maintain his employer's advantage over competitors. Both of these constraints violate the rather strong scientific norms that underlie and form the basis of the invisible college. The first of these norms demands that science be free to choose its own problems and that the community of colleagues be the only judges of the relative importance of possible areas of investigation, and the second is that the substantive findings of research are to be fully assigned and communicated to the entire research community. The industrial organization, by preventing its employers from adhering to these two norms, impedes the formation by technologists of anything resembling an invisible college.

Impact of "Localism" on Communication

What is the effect of this enforced "localism" on the communication patterns of engineers? Because proprietary information must be protected to preserve the firm's position in a highly competitive marketplace, free communication among engineers of different organizations is greatly inhibited. It is always amusing to observe engineers from different companies interacting in the hallways and cocktail lounges at conventions of professional engineering societies. Each one is trying to draw the maximum amount of information from his competitors while giving up as little as possible of his own information in return. Often the winner in this bargaining situation is the person with the strongest physical constitution.

Another result of the concern over divulging proprietary information will be observed in looking at an engineer's reading habits. A good proportion of the truly important information generated in an industrial laboratory cannot be published in the open literature because it is considered proprietary and must be protected. It is, however, published within the organization, and, for this reason, the informal documentation system of his parent organization is an extremely important source of information for the engineer.

The Effect of Turnover

It is this author's suspicion that much of the proprietary protectionism in industry is far overplayed. Despite all of the organizational efforts to prevent it, the state of the art in a technology propagates quite rapidly. Either there are too many martinis consumed at engineering conventions or some other mechanism is at work. This other mechanism may well be the itinerant engineer, who passes through quite a number of organizations over the course of a career. Shapero (1968) makes this point very strongly and musters evidence in support of it. He points to the finding by the Engineering Manpower Commission of the Engineers' Joint Council that the turnover in industries classified as "aircraft and parts," "communications," "electrical-electronics," "instruments," and "R&D" is 12.1 percent. He goes on to say that his own "limited data . . . indicate that the turnover rates of [engineers and scientists] are considerably higher than 12.1 percent in many defense R&D establishments." Turnover, both voluntary and as a result of layoffs, has at times certainly been far in excess of 12 percent. And for certain years in individual firms it has been far more than 25 percent. One has merely to recall the situation on Boston's Route 128, in the Seattle area, or in Los Angeles over recent years to realize the levels that turnover can periodically reach.

Each time that an engineer leaves an employer, voluntarily or otherwise, he carries some knowledge of the employer's operations, experience, and current technology with him. We are gradually coming to realize that human beings are the most effective carriers of information and that the best way to transfer information between organizations or social systems is to physically transfer a human carrier. Roberts's studies (Roberts and Wainer, 1967) marshal impressive evidence for the effective transfer of space technology from quasi-academic institutions to the industrial sector and eventually to commercial application in those instances in which technologists left university laboratories to establish their own businesses. This finding is especially impressive in view of the general failure to find evidence of successful transfer of space technology by any other mechanism, despite the fact that many techniques have been tried and a substantial amount of money has been invested in promoting the transfer.

This certainly makes sense. Ideas have no real existence outside of the minds of men. Ideas can be represented in verbal or graphic form, but such representation is necessarily incomplete and cannot be easily structured to fit new situations. The human brain has a capacity for flexibly restructuring information in a manner that has never been approached by even the most sophisticated computer programs. For truly effective transfer of technical information, we must make use of this human ability to recode and restructure information so that it fits into new contexts and situations. Consequently, the best way to transfer technical information is to move a human carrier. The high turnover among engineers results in a heavy migration from organization to organization and is therefore a very effective mechanism for disseminating technology throughout an industry and often to other industries. Every time an engineer changes jobs he brings with him a record of his experiences on the former job and a great amount of what his former organization considers "proprietary" information. Now, of course, the information is usually quite perishable, and its value decays rapidly with time. But a continual flow of engineers among the firms of an industry ensures that no single firm is very far behind in knowledge of what its competitors are doing. So the mere existence of high turnover among R&D personnel vitiates much of the protectionism accorded proprietary information.

As for turnover itself, it is well known that most organizations attempt to minimize it. If all of the above is even partially true, a low level of turnover could be seriously damaging to the interests of the organization. Actually, however, quite the opposite is true. A certain amount of turnover may be not only desirable but absolutely essential to the survival of a technical organization, although just what the optimum turnover level is for an organization is a question that remains to be answered. It will vary from one situation to the next and is highly dependent upon the rate at which the organization's technical staff is growing. After all, it is the influx of new engineers that is most beneficial to the organization, not the exodus of old ones. When growth rate is high, turnover can be low. An organization that is not growing should welcome or encourage turnover. The Engineers' Joint Council figure of 12 percent may even be below the optimum for some organizations. Despite the costs of hiring and processing new personnel, an organization might desire an even higher level of turnover. Although it is impossible to place a price tag on the new state-of-the-art information that is brought in by new employees, it may very well more than counterbalance the costs of hiring. This would be true at least to the point where turnover becomes disruptive to the morale and functioning of the organization.

COMMUNICATION PATTERNS IN SCIENCE AND TECHNOLOGY

The difference between communication patterns in science and technology is amply illustrated by a comparison of several of the R&D projects in this study. Among the sample of nineteen parallel projects, seventeen were clearly developmental in their nature. The remaining two had clear-cut goals, but these were directed toward an increased understanding of a particular set of phenomena. While the information generated by these two teams would eventually be used to develop new hardware, this was not the immediate goal of the teams, who were far more interested in the phenomena than in the application. For this reason, their work can be considered to be much more scientific than technological in nature.

A comparison of the two scientific projects with the seventeen technological projects (table 3.2) shows a marked disparity in the use of eight information channels. The scientists engaged in the phenomena-oriented project concentrated their attention heavily

Table 3.2 Sources of Messages Resulting in Technical Ideas Considered
During the Course of Nineteen Projects

Channel	Seventeen Technological Projects		Two Scientific Research Projects	
	Number of Messages Produced	Percentage of Total	Number of Messages Produced	Percentage of Total
Literature	53	8	18	51
Vendors	101	14	0	0
Customer	132	19	0	0
Other sources external to the laboratory	67	9	5	14
Laboratory technical staff	44	6	1	3
Company research programs	37	5	1	3
Analysis and experimentation	216	31	3	9
Previous personal experience	56	8	7	20

upon the literature and upon colleagues outside their laboratory
organization. The engineers spread their attention more evenly
over the channels and received ideas from two sources unused by
the scientists. The customer (in this case, a government laboratory) suggested a substantial number of ideas, demonstrating the
importance of the marketplace for technologists. Vendors are
another important channel in technology because they are important potential suppliers of components or subsystems, and they
provide information that they hope will stimulate future business.
Involvement in the marketplace, either through the customer or
potential vendors, exerts a significant influence upon the communication system, providing channels for the exchange of information in two directions and connecting buyers and sellers through
both the procurement and marketing functions of the organization.

The extent to which scientists and engineers differ in the degree
to which they use oral and written channels can be seen even
more clearly in comparing the way in which they allocate their
time (table 3.3). A comparison of time spent using literature or

Table 3.3 A Comparison of Engineers' and Scientists' Allocation of Time in Communication

Channel	Proportion of time spent during:	
	Twelve Technological Problems	Two Scientific Projects
Literature use	7.9%	18.2%
Total time in communication	16.4	28.6
Total time reported (man-hours)	20,185	1,580

talking by the engineers working on twelve technological projects and the scientists of the two phenomena-oriented research projects shows the scientists devoting 75 percent more of their time to communication. Most of this increased communication time is given to the literature. Both the engineers and scientists spent about 10 percent of their overall time discussing technical matters with colleagues, but the engineers spent more time in personal contact than in reading.

This comparison is quite revealing. Despite all the discussion of informal contact and invisible colleges among scientists (and scientists do make extensive use of personal contacts), it is the engineer who is more dependent upon colleagues. The difference between communication behavior of scientists and engineers is not simply quantitative, however. The persons contacted by scientists are very different from those contacted by engineers, and the relationship between the engineer and those with whom he communicates is vastly different from the relationship that exists among scientists. In written channels, too, there are significant differences. The literature used by scientists differs qualitatively from that used by engineers. And the engineer not only reads different journals, but, as we shall see in the next chapter, he uses the literature for entirely different purposes.

THE RELATION BETWEEN SCIENCE AND TECHNOLOGY
Given the vast differences between science and technology, how

do the two relate to each other? This is a question that has intrigued a number of researchers in recent years. It is generally assumed that the two are in some way related, and in fact national financing of scientific activity is normally justified on the basis of its eventual benefits to technology. Is there any basis for this, and what, if any, is the relation of science to technology? How are the results of scientific activity incorporated into technological developments? To what extent is technology dependent upon science? What are the time lags involved?

The Process of Normal Science

Kuhn (1962) describes three classes of problems that are normally undertaken in science:

1. The determination of significant facts that the research paradigm has shown to be particularly revealing of the nature of things.[2]

2. The determination of facts, which (in contrast with problems of the first class) may, themselves, be of little interest, but which can be compared directly with predictions made by the research paradigm.

3. Empirical work undertaken to articulate the paradigm theory. The first two of these—the precise determination and extension to other situations of facts and constants that the paradigm especially values (for example, stellar position and magnitude, specific gravities, wave lengths, boiling points) since they have been used in solving paradigmatic problems, and the test of hypotheses derived from the central body of theory—will not concern us here. These are the normally accepted concerns of science, but the third-listed function is probably the most important, and I shall address myself to this category of activity that comprises empirical work undertaken to extend and complete the central body of theory. It may, itself, be subdivided into three classes of activity (Kuhn, 1970):[2]

1. The determination of physical constants (gravitational constants; Avogadro's Number; Joule's Coefficient; etc.).

2. The development of quantitative laws. (Boyle's, Coulomb's, and Ohm's Laws).
3. Experiments designed to choose among alternative ways to applying the paradigm to new areas of interest.

Within the third class lie problems that have resulted from difficulties encountered during the course of scientific research or during the process of technological advance. This, as we shall see, is a form of scientific activity of extreme interest and importance.

The Dependence of Technology on Science

Despite the long-held belief in a continuous progression from basic research through applied research to development, empirical investigation has found little support for such a situation. It is becoming generally accepted that technology builds upon itself and advances quite independently of any link with the scientific frontier, and often without any necessity for an understanding of the basic science which underlies it. Price (1965b), a strong advocate of this position, cites Toynbee's view that

physical science and industrialism may be conceived as a pair of dancers, both of whom know their steps and have an ear for the rhythm of the music. If the partner who has been leading chooses to change parts and to follow instead, there is perhaps no reason to expect that he will dance less correctly than before.

Price goes on to marshal evidence refuting the idea of technology as something "growing out of" science and to make the claim that communication between the two is at best a "weak interaction." Communication between the two is restricted almost completely to that which takes place through the process of education.

   Singer (1959), in setting the scene for his history of scientific thought, describes science as the activity or process of knowledge making. He stresses that "science . . . is no static body of knowledge but rather an active process that can be followed through the ages." It is a stream of human activity devoted to building a store of knowledge and can be traced back to the beginning of recorded history. Science can thereby be represented as a stream of events over time cumulating in a body of knowledge. There are two other streams of human activity that operate parallel to science and that function both as contributors to scientific development and as beneficiaries of scientific accomplishment. First there is the activity we have labeled "technology." This is a stream of human activity oriented toward incorporating human knowledge into physical hardware, which will eventually meet with some human use. Then there is a much more general form of human activity in which the ideas of science and the hardware of technology are actually put to some use in the stream of human affairs. This last stream we will label *utilization* (figure 3.1).

The activities of technology and of utilization in commerce, industry, welfare, and war, while at various times in close harness with science, have developed for the most part independently. Science builds on prior science; technology builds on prior technology; and utilization grows and spreads in response to needs and benefits.

The familiar notion of science providing the basis upon which technology is built to be later utilized in commerce or industry has been shown by the historians of science to have only a limited basis in historical fact. Civilizations have often emphasized activity in one or two of these areas to the exclusion of the others. The Greeks, for example, were very active in science, but they were relatively little concerned with the practical applications or implications of their discoveries. The Romans, in contrast, developed a highly practical civilization, which was greatly concerned with the building of artifices to aid in coping with the physical and social environment. They devoted much effort to the construction of roads and aqueducts and of improvement of armor and weapons without much concurrent increase in their understanding of the natural basis of their developments. History shows quite independent paths through the succeeding centuries to the present time. The three streams appear now in rapid parallel growth;

Figure 3.1 Science, Technology, and the Utilization of Their Products, Showing the Normal Progression from One to the Other

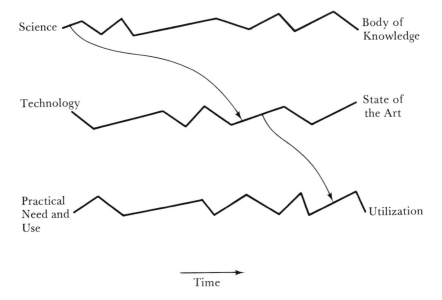

an increased emphasis in one is usually accompanied by an increase in the other two. It is probable that the streams are more closely coupled now than they have been historically, but the delays encountered in any of the communication paths between them remain substantial.

The Flow of Information Between Science and Technology

Over the past ten years several studies have attempted to trace the flow of information from science to technology. In one of the earlier of these, Price (1965b), after investigating citation patterns in both scientific and technological journals, concluded that science and technology progress quite independently of one another. Technology, in this sense, builds upon its own prior developments and advances in a manner independent of any link with the current scientific frontier and often without any necessity for an understanding of the basic science underlying it.

Price's hypothesis certainly appears valid in light of more recent evidence. There is little support for direct communication between science and technology. The two do advance quite independently, and much of technology develops without a complete understanding of the science upon which it is built.

Project Hindsight was the first of a series of attempts to trace technological advances back to their scientific origins. Within the twenty-year horizon of its backward search, Hindsight was able to find very little contribution from basic science (Sherwin and Isenson, 1967). In most cases, the trail ran cold before reaching any activity that could be considered basic research. In Isenson's words, "It would appear that most advances in the technological state of the art are based on no more recent advances than Ohm's Law or Maxwell's equations."

Project TRACES (IIT Research Institute, 1968), partially in response to the Hindsight results, succeeded in tracing the origins of six technological innovations back to the underlying basic sciences but only after extending the time horizon well beyond twenty years. In a follow-up, Battelle (1973) investigators found similar lags in five more innovations. In yet another recent study, Langrish found little support for a strong science-technology interaction. Langrish wisely avoided the problem of differentiating science from technology. He categorized research by the type of institution in which it was conducted—industry, university, or government establishment. In tracing eighty-four award-winning innovations to their origins, he found that "the role of university as a source of ideas for [industrial] innovation is fairly small" (Langrish, 1971) and that "university science and industrial technology are two quite separate activities which occasionally come into contact with each other" (Langrish, 1969). He argued very strongly that most university basic research is totally irrelevant to societal needs and can be only partially justified for its contributions through training of students.

Gibbons and Johnston (1973) attempted to refute the Langrish hypothesis. They presented data from thirty relatively small-scale technological advances and found that approximately one-sixth of the information needed in problem solving came from scientific sources. Furthermore, they claimed greater currency in the scientific information that was used. The mean age of the scientific journals they cited was 12.2 years. This is not quite twenty, but with publication lags, it can safely be concluded that the work was fourteen or fifteen years old at the time of use. They showed considerable use of personal contact with university scientists, but nearly half of these were for the purpose of either referral to other sources of information or to determine the existence of specialized facilities or services. So, while Gibbons and Johnston may raise some doubt over the Price-Langrish hypothesis, the contrary evidence is hardly compelling.

The evidence, in fact, is very convincing that the normal path from science to technology is, at best, one that requires a great amount of time. There are certainly very long delays in the system, but it should not be assumed that the delays are always necessarily there. Occasionally, technology is forced to forfeit some of its independence. This happens when its advance is impeded by a lack of understanding of the scientific basis of the phenomena with which it is dealing. The call then goes out for help. Often a very interesting basic research problem can result, and scientists can be attracted to it. In this way, science often discovers voids in its knowledge of areas that have long since been bypassed by the research front. Science must, so to speak, backtrack a bit and increase its understanding of an area previously bypassed or neglected.

Morton (1965) described several examples in which technology has defined important problems for scientific investigation. He pointed out that progress in electron tube technology at one time appeared to have reached an upper limit of a few megacycles in frequency response. With the rapidly increasing amount of radio frequency communication, it was clearly desirable to extend the range of usable frequencies above that limit to increase the number of channels available to communicators and to allow the use of larger band widths, thereby increasing the amount of information per unit of time that could be transmitted. This difficulty forced the realization that electron tube technology had advanced without a real understanding of the principles involved. It did this largely by "cut and try" methods, manipulating the geometry of the elements and the composition of the cathode materials with little real understanding of the fundamental physics underlying the results. This block to the advance of a burgeoning technology forced a return to basic classical physics and a more detailed study of the interactions of free electrons and electromagnetic waves. The return allowed scientists to fill a gap in their understanding and subsequently permitted the development of such microwave amplifiers as the magnetron, klystron, and traveling wave tube. To quote Morton (1965, p. 64), "The important lesson to be learned from the vigorous past of electronics is not that it was always close to the *new frontiers* of basic science. Rather, it was the conscious or unconscious recognition of *relevant* physical phenomenon and materials, *old or new* which could fulfill a critical need or break an anticipated technological barrier."

Note that there was first of all communication of a problem from technology to science, followed by a relatively easy transfer of scientific results back to the technologists. The two conditions are clearly related. When technology is the source of the problem, technologists are ready and capable of understanding the solution and putting it to work.

Additional support for this idea is provided by Project Hindsight (Sherwin and Isenson, 1967). While in most cases, Hindsight was unable to find any contribution to technology from basic science, it is the exception to this discontinuity between science and technology that chiefly concerns us at the present. Isenson reports[3] that he discovered exceptions to his general finding and that these exceptions are usually characterized by a situation in which, similar to Morton, technology has advanced to a limit at which an understanding is required of the basic physical science involved. Thus technology defines a problem for science. When this problem is attacked and resolved by scientists, its solution is passed immediately into technology. A close coupling thus exists for at least an isolated point in time, and the researchers of Project Hindsight were able to trace the record back from an improved system in what we have labeled the "utilization stream" through an advance in the technological state of the art to the closure of a gap in the body of scientific knowledge. To distinguish this latter form of research from "frontier science," I propose calling it "gap-filling science."

Gap-filling science is by its nature directly responsive to technological need, and the advance of technology is often contingent upon the pursuit of gap-filling science. So when the connection between science and technology is of this form, little delay is encountered in the transfer of information (figure 3.2). Communication is rapid and direct, and the long delays of the normal transfer process are circumvented. The transfer from gap-filling science can be further accelerated by including in the technological development team former scientists or individuals whose training was in science. The advantages of such a strategy were clearly demonstrated

Figure 3.2 Science, Technology, and the Utilization of Their Products, Showing Communication Paths Among the Three Streams. (a) The normal process of assimilation of scientific results into technology. (b) Recognized need for a device, technique, or scientific understanding. (c) The normal process of adoption of technology for use. (d) Technological need for understanding of physical phenomena and its response (from Marquis and Allen, 1966).

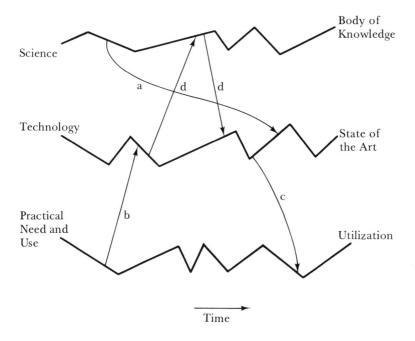

Table 3.4 Citations From Engineering Journals to Other Technological and
Scientific Journals

	Proportion of references[a] citing:			
	Technological journals	Scientific journals	Ratio of technological to scientific citations	Total number of citations
Journal of Applied Mechanics[b]	38.7%	6.4%	6.05	1,172
Burton's (1959b) data for mechanical engineering journals[c]	62.0	19.7	3.15	1,278
Burton's (1959a) data for chemical engineering journals[c]	55.1	27.4	2.01	1,741
Burton's (1959c) data for metallurgical engineering journals[c]	62.4	31.8	1.96	2,639
IEEE Transactions on Electron Devices[d]	43.4	25.0	1.74	865
Nuclear Science and Engineering[e]	18.3	19.8	.93	1,669

[a]References to books, unpublished reports, and unclassified journals constitute
the remainder.

[b]January through December 1965.

[c]January through December 1957. The values shown result from a reanalysis
of the original data. Proportions shown for the Burton data are based on a
count which does not include books or unpublished reports. Hence, they
are somewhat larger in absolute magnitude compared with the other proportions; thus the ratios are the only comparable figures.

[d]January through October 1965.

[e]January through December 1965.

From Marquis and Allen (1966)

during World War II when many scientists became engineers, at
least temporarily, and were very effective in implementing the
results of fundamental research.

The point to be made is that at least a segment of basic science is
not conducted at what is called the "frontier" of knowledge.
Technology—and often investigation in a different scientific area—
will raise problems that attract investigators to an area that has been

worked on before. The investigation then proceeds, looking perhaps from a somewhat different vantage at items that had not previously been deemed important phenomena. That such investigations are searching in what had been considered secure territory makes them no less fundamental in their nature. To draw upon the National Science Foundation's definition of basic research, because these researchers are being directed back over old ground does not mean that their primary aim cannot be "a fuller knowledge or understanding of the subject under study, rather than a practical application" (National Science Foundation, 1965).

A second element can be added to Price's hypothesis. While technology and science in general may progress quite independently of each other, there very probably are some technologies that are more closely connected with science than others. For example, electronics technology is more closely related to frontier work in physics than say, mechanical technology. Nuclear technology should be more closely coupled to the advance of physical knowledge than either of these two. There is some evidence to support this variance in the nature of the coupling. The engineering fields examined in table 3.4 show the ratio of technological to scientific citations to range from about six to one to less than one to one. The data clearly indicate that a wide variation exists in the degree to which technologies are coupled to their respective sciences.

NOTES

1. Even Pelz and Andrews (1966), who are careful to preserve the distinction throughout most of their book, a study of 1,300 engineers and scientists, chose *Scientists in Organizations* as its title, forgetting the majority of their sample.

2. Kuhn's term "research paradigm" refers to a body of scientific theory and evidence whose "achievement [is] sufficiently unprecedented to attract an enduring group of adherents away from competing modes. Simultaneously, it [is] sufficiently open ended to leave all sorts of problems for . . . practitioners to resolve."

3. Personal communication, April 22, 1966.

4 THE TECHNOLOGICAL LITERATURE

Turning now to the problems of communication in technology, let us examine the way in which the technologists on twenty-two preliminary design projects obtained the technical information necessary to complete their tasks. The twenty-two projects consist of nine matched sets of parallel or "twin" projects (one was a triplet) and three unmatched single projects. The unmatched projects were originally members of parallel pairs, but their parallel partners either refused to cooperate or supplied data of such poor quality that they had to be discarded.

Data were gathered from the twenty-two projects by means of solution development records, time allocation forms, and interviews conducted either at convenient points during the project or following project completion. Reliable data from solution development records were obtained from seventeen of the twenty-two projects, all of which belonged to matched sets. Time allocation data were obtained from twelve of the twenty-two projects, including all three of the unmatched single projects and four and one-half of the pairs. Performance comparisons were therefore possible using either the message count approach of the solution development record or the amount of time devoted to information activities as reported on the time allocation forms.

ALLOCATION OF TIME TO LITERATURE USE
In considering the time allocation data, the reader should bear in mind that all of the projects were preliminary design studies and only three of the problems involved even prototype hardware production. Consequently, the number of people assigned to a project was small (median = 6), and the proportion of time spent in communication and analysis may be high compared to that which might be spent by an engineer involved in later stages of hardware development. As a project nears the actual production of hardware, an engineer would be expected to spend more time in setting up and conducting systems tests and in negotiating the actual manufacture, and he would have proportionately less time for information gathering and analysis. The overall level of effort

allocated to information processing is therefore a bit higher than would probably be found later in the development process, but comparison of these data with the results of other studies of development work shows remarkable agreement in the proportions of time allotted to different forms of information processing. As a result, it is probably safe to assume that little distortion is introduced by the decision to study the preliminary design phase of technology.

One other point must be made concerning the sample. Sometimes it was impossible to obtain time allocation data from all of the engineers who worked on a project. The total number of man-hours lost as a result of this is probably not very great; we estimate it to be less than 10 percent of the total time reported. Often people were assigned for short periods of time during the project, and it was impossible to keep track of these people and obtain forms from them. In addition, for one of the four parallel pairs, time data were gathered from only about half of the people assigned to the project in each organization (the other half submitted solution development records).[1] This sample was not arrived at randomly but comprised the people who remained after lead engineers were selected to complete solution development records. As a result of this, the engineers who supplied time allocation data were generally not the people principally responsible for subsystems but were those at the bottom of the project hierarchy. This, too, may cause some inflation of the figures for the proportion of time spent in information gathering and processing, but this time the bias should be toward a more realistic appraisal of how the typical "bench" engineer spends his time.

Analysis of the time allocation data shows that almost 95 percent of the engineers' time was given to gathering and processing technical information (table 4.1).[2] The gathering of information from outside the project team accounts for about 16 percent of the engineers' time, and is almost evenly divided between written sources and people.

Eight of the twelve projects in table 4.1 were paired in four sets.

Table 4.1 Comparison of Time Allocation Among Four Activities by Higher-
and Lower-Rated Project Teams

	Percent of Total Time Allocated		
	Total for Twelve Development Projects	Four Higher-Rated Projects	Four Lower-Rated Projects
Analysis and experimentation	77.3	77.9	71.4
Literature use	7.9	5.0	5.3
All communication (including literature)	16.4	13.9	13.4
Other activity	6.3	8.2	15.7
Total time reported (man-hours)	20,185	6,566	7,975

For these four pairs performance evaluations were obtained from
competent technical personnel in government laboratories spon-
soring the projects. The evaluations were provided on an informal,
confidential basis and were directed to only the technical quality
of the problem solutions. Comparison of the communication
patterns of the high and low performing teams (table 4.1) shows
very little difference: both high and low performers[3] spent about
5 percent of their total time reading.[4] The very slight difference
in proportion between high and low performances is obviously
not statistically significant, and any difference in overall performance on the projects certainly cannot be attributed to the
extent to which the participants read the technical literature.

The failure to find any relation between project performance
and time spent reading technical literature may at first seem
strange in light of the traditional image of the literature as the
principal vehicle for transmitting technical information. It must
be remembered that we are now dealing with technology rather
than science, and the literature plays a very different role in
technology. To learn more about this role one must probe a bit
deeper into the way in which literature is used. For example, the
results in table 4.1 are averaged over the entire length of a project.

It is certainly reasonable to expect the amount of time spent in reading to vary considerably over this span. One would expect heavy use of the literature at the outset of a project, with a gradual decline as the project approached completion. This is in accord with the assumed custom of conducting a literature search at the outset of a project and then returning to the literature only for specific problems later on.[5]

Data from the four sets of twin projects adhere quite closely to the expected pattern (figure 4.1). There was an initial surge, evidenced especially by the lower performing teams, followed by a gradual decay toward project completion. Nearly half of the total requirement for literature occurred in the first third of the project. It is not clear why the poorer performers spent more time using the literature at the start of a project. Perhaps they were less well prepared or experienced in the area and attempted to offset this deficit by an extensive literature search. On the other hand, it is entirely conceivable that the high-performing teams had completed this heavy reading phase prior to formal initiation of the project. Unfortunately, with the available data we are unable to test this possibility. Once the first one-third of the project was completed, there was little difference between high and low performers. Both tapered off their use of literature as the project progressed.

The implications of this time dependency are readily apparent. The point at which to make a special effort to supply a project with state-of-the-art literature is during the first third of its life. From that point on literature becomes far less important. Certain kinds of documents and reports will always be valuable, but the prime need for literature is in the first few weeks of a project. If special library assistance or the services of a retrieval system are to be provided, these might be terminated or reduced relatively early in the project. Information specialists are most needed during the critical early weeks of a project, but their commitment of time to the project can be substantially reduced once the first one-third is completed. The allocation of an information specialist's time

Figure 4.1 Comparison of Time Spent by Higher- and Lower-Performing Project Teams in Literature Use. Calculating average proportions for the six periods will produce figures slightly different from those in table 4.1 because the number of man-hours on which proportions are based vary from period to period.

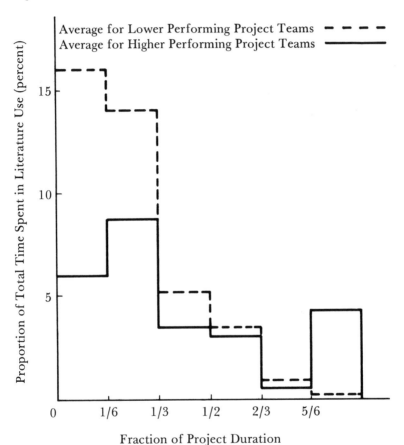

Fraction of Project Duration

should not be evenly distributed over all projects in a laboratory but should be made a function of the phase spectrum of projects in the laboratory.

LITERATURE AS AN IDEA SOURCE

Of all the functions performed by any information source, probably the most important is the generation of new ideas. The development of new ideas or new solutions to problems is a creative process in which the individual should be exposed by the information system to a very broad range of possibilities. There is some evidence to show that during very early phases of problem solving, it is beneficial to consider a large number of possible solutions to a problem (Allen and Marquis, 1963; Allen, 1966a; Osborne, 1957). In the examples shown in chapter 2, the engineers each considered a number of possible solutions to their problems. Some of these were developed at the outset of the project, and others were generated after work was underway. Possible solutions were reported on solution development records. As new ideas appeared on the records, the engineers were contracted to determine the sources that were responsible for producing the idea.

About 11 percent of all the idea-generating "messages" were obtained through literature (figure 4.2). Nearly all of the remaining messages came through personal contacts. The only possible exception to this is under the category labeled "company research." Results of other research projects in the firm were often communicated simultaneously by written and oral means, and it was sometimes difficult to determine which of the two was operating in a given situation. Some of the company report literature is therefore not included under the category "literature."[6] This is the reason that "company research" is shown separately in figure 4.2. In the extreme, however, even if all of the results of company research were transmitted by written means, it would add only another 7.5 percent, producing an unimpressive total of 18.5 percent of all idea-generating messages. Although "customer" is not a written source of information, it is shown in figure 4.2

Figure 4.2 The Sources of 494 Messages Resulting in Ideas Considered as Potential Solutions to Technological Problems (Data from Seventeen Projects)

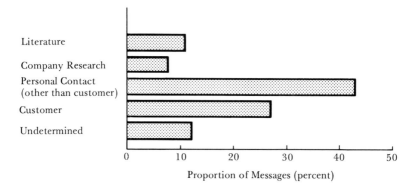

because it produced so many ideas and because heavy customer involvement is probably unique to those situations in which there is a well-organized and technically sophisticated customer. The role of the customer will be less active when he lacks these characteristics.

USE OF THE LITERATURE IN PROBLEM DEFINITION

Literature can, of course, be used for functions other then generating ideas. There are a number of well-defined stages that a technologist must go through in defining a problem he is working on (Allen and Frischmuth, 1969). The use of literature is compared with that of personal contact, for three of these stages or functions, in figure 4.3. Problem definition for engineering problems is often both more important and more difficult than idea generation. Problem definition is particularly critical, both because it is at this point that "need-oriented" (Utterback, 1975) information enters the system and because solution quality is so highly dependent upon proper assessment and structuring of the problem (Wertheimer, 1959).

In the problem definition phase, as well as idea generation, personal contacts are used far more than literature to generate information. Aggregating across all three functions shows that personal contact provided more than five times the number of messages supplied by written sources.

Thus far in the analysis, measurements have been aggregated for the entire project; they will now be broken down to an individual level. Analyzing the data at this level introduces a possibility of offsetting differences that cloud the variations among individual project members. There can be great variation among the individuals working on a project, as well as among the types of problems encountered. Furthermore, even the most successful projects may have subsystems whose design is not up to par. When we say that one of the project teams was chosen as the higher performer, it does not necessarily mean that all of the subsystem designs it submitted were evaluated as better than the competitor's.

Figure 4.3 Relation of Problem-Solving Function to Information Channel
Use (Four Technological Projects)

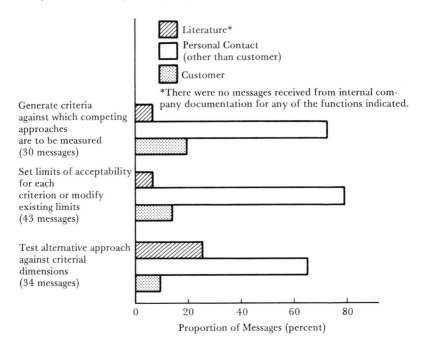

Proportion of Messages (percent)

(Presumably the majority or the most important ones were.)
An example will show how this can lead to difficulties when analyzing at the project level. On a project with three subsystems, the
effort in literature search might be allocated as shown in table
4.2. If team A turns in the better performance on subsystems 1 and
3 while B performs better on 2 and the subsystems are weighted
equally in determining the overall project evaluation, then team A
will be considered the high performer. Both teams will have spent
the same amount of time in literature search so there will appear to
be no difference between high and low performing teams. Yet at a
subsystem level, only those subsystem designs for which a greater
amount of time was spent in literature search received high evaluation. For this reason, it is best to obtain evaluations at the level
at which the independent variable itself operates. Literature search
is not performed by project teams; it is performed by individual
engineers. When performance is evaluated at the project team
level, only the total or mean level of literature use can be compared with it. This ignores the variance among the individual
engineers on the team. A minority of high-performing engineers on
a low-performing team may be using the literature to an even
greater extent than all of the members of the high-performing
team. This could result in a conclusion that the low-performing
team spent less time with literature, when in fact, on a subsystem
level, just the opposite were true. This difficulty can be overcome
by shifting the analysis down to the subsystem level. While this
could not be done with the time allocation data, it can be for
idea generation. The information sources responsible for each of
the subsystem solutions are known, as are the relative evaluations
of each of these solutions.

SOURCES OF IDEAS AND PERFORMANCE
Eighty pairs of subproblems were studied. Technical monitors in
the government laboratories were able to say, with confidence,
that one solution was technically superior to the other for twenty-
five of these pairs. In the remaining fifty-five cases, evaluations

Table 4.2 Time Spent in Literature Search (in Man-Hours)

Subsystem	Team A	Team B	Team with Highest Evaluation
1	10	5	A
2	0	10	B
3	10	5	A
Total	20	20	—

were either tied or no evaluation was available. Division between superior and poorer solutions allows a number of comparisons to be made; for example, the sources responsible for generating superior ideas can be compared with those producing poorer ideas.

When this is done, it appears at first that written material tended to produce poorer solutions (figure 4.4). This result is not statistically reliable, however, and the strongest statement that can be made is that literature produced neither more nor fewer high-rated solutions. There is no relation between solution quality and use of written sources. This is not the first time that we have reached this conclusion. An earlier study of twenty-two R&D proposal competitors (Allen, 1964) found no correlation between the extent to which 156 proposal teams made use of the technical literature and the quality of their proposals.

There are a number of possible reasons (besides the obvious one that the literature makes no contribution to performance) that could explain the situation. First of all, even though the average engineer does not use the literature very heavily, there is a possibility that some nonaverage engineers do. Merely looking at the mean of the distribution does not tell the complete story. There may well be a minority of engineers who are heavy literature users and who are able to contribute to project performance in some way that we have not yet ascertained. This is a question to which we shall turn in chapter 6.

As a second consideration, there are certainly many kinds of literature. Some of these may make a strong contribution to

Figure 4.4 Solution Quality as a Function of Information Source (Literature)

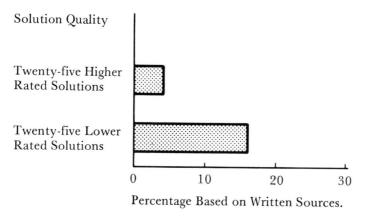

Percentage Based on Written Sources.

performance, only to be outweighed in our analysis by other written media that contribute nothing. A more detailed analysis of specific literature sources (books, journals, trade magazines, government publications, and so on) will be necessary to address this possibility. Although we are not able in this study to determine the variations in performance of these different media, there is evidence of the extent to which each is used.

A MORE DETAILED EXAMINATION OF WRITTEN MEDIA

On twelve of the projects, participants were asked to answer three questions about each item of written material that they used.[7] These questions were addressed to the following points.

1. Identity. The name of the book or journal or the title and organizational sponsorship of unpublished reports.
2. Means of Acquisition. The method by which the material came into the hands of the user.
3. Function. The specific problem-solving function for which the material was used.

Looking first at the identity of the publications that were read, there are two major categories of publication that engineers use. The first of these might be called formal literature. It comprises books, professional journals, trade publications, and other media that are normally available to the public and have few, if any, restrictions on their distribution. Informal publications, on the other hand, are published by organizations usually for their own internal use; they often contain proprietary material and for that reason are given a very limited distribution. On the average, engineers divide their attention between the two media on about an equal basis, only slightly favoring the informal publications (table 4.3). Because engineering reports are usually much longer than journal articles and because books are used only very briefly for quite specific purposes, each instance of report reading takes twice as long as an instance of journal or book reading. The net result is a threefold greater expenditure of time on informal reports. We can conclude from this brief overview that the

Table 4.3 Relative Use of Formal and Informal Literature by Engineers

	Percentage of Instances	Percentage of Total Man-Hours	Average Number of Man-Hours per Instance
Formal literature (books, journals, periodicals)	45	26	1.5
Informal literature (unpublished reports)	55	74	3.2

unpublished engineering report occupies a position that is at least as important as that of the book or journal in the average engineer's reading portfolio. Let us turn now and consider formal and information channels, separately.

THE FORMAL ENGINEERING LITERATURE

Textbooks are the most frequently encountered of all formal publications (figure 4.5). Furthermore, they account for the greatest amount of time expended (31 percent). By both measures, textbooks make up almost one-third of the literature used by engineers. Trade journals and controlled circulation journals (those published by an organization that is not a professional society, for which there is normally no subscription charge provided the subscriber belongs to a qualified audience, and which are entirely supported by advertising) follow in second and third place, respectively. Trade journals and controlled circulation journals are each responsible for about 20 percent of the instances of use. There are very few references to scientific journals, and even the professionally sponsored engineering journals are little used. This last category (chiefly the *Proceedings of the IEEE, AIAA Journal,* and *Mechanical Engineering*) accounts for only 11 percent of the instances in which literature was used and 10 percent of the total reading time. Although frequency of use is all that is shown in figure 4.5, the number of man-hours spent in each instance was also measured. Since both measures correlate almost perfectly, only one is shown. The more frequently a

Figure 4.5 Use of Formal Literature Sources by Engineers on Twelve Development Projects (Based on 203 Instances of Use). The category "other journals" comprises four references in which journal title was not indicated and five references to *Scientific American*.

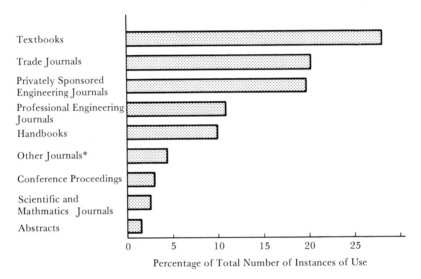

Percentage of Total Number of Instances of Use

particular form of publication is used, the greater the amount of time that is devoted to it. The mean duration of each instance of use is fairly constant across all publication forms and is about one and one-half hours. There is naturally a large variance in the duration of instances. Some took five minutes; others required several hours.

A number of other studies (Herner, 1954; Shaw, 1956; Fishendon, 1959; Scott and Williams, 1959; Berul et al., 1965) show a similar distribution of channel usage. Seldom in any of these studies do the formal literature channels compare in degree of use with the informal channel of unpublished reports. Even among the formal channels, textbooks, profit-oriented engineering journals, and trade magazines head the list, with the professional society publications far behind. Further, the wide variety of populations that were studied indicates that this low use of the professional literature is not a phenomenon that is limited to any specialized segment of the engineering profession. Essentially the same situation holds for aerospace engineers, British electronics engineers, nuclear technologists, and engineers employed by the U. S. Forest Service. The publications of the professional engineering societies in all of these diverse fields are little used by their intended audience.

Why should this be so? The answer is not difficult to find. Most professional engineering journals are utterly incomprehensible to the average engineer. They often rely heavily upon mathematical presentations, which can be understood by only a limited audience. The average engineer has been away from the university for a number of years and has usually allowed his mathematical skills to degenerate. Even if he understood the mathematics at one time, it is unlikely that he can now. The articles, even in engineering society journals, are written for a very limited audience, usually those few at the very forefront of a technology. Just as in science, the goal of the author is not to communicate to the outsider but to gain for himself the recognition of his peers. No attempt is made to interpret or translate the material into a form that the average practicing engineer can understand and use. The

publications of the Institute of Electrical and Electronics En-
ineers provide a perfect example of this. The *Proceedings* of this
society are utterly incomprehensible to most electrical or electronics engineers and the various *Transactions* are only slightly so.
The practicing engineer must look elsewhere for information, and
he goes either to informal sources or to any of a number of
journals that have arisen to fill the gap between the socicty journals
and the average engineer's capabilities. The latter have grown up,
for the most part, outside the surveillance of the engineering societies and are published by private organizations whose purpose
is to gain a profit through sale of advertising space.

What is wrong with this situation? Perhaps nothing. Some of the
profit-making journals are quite good. The problem lies in the
fact that there is no guarantee of quality or even honesty in their
presentations. Quality control is solely in the hands of the editors.
And since all such journals are supported mainly by advertising
(subscriptions are often free to anyone who can qualify as being
a member of the audience whom the advertisers hope to reach),
there must exist a rather strong aversion to rejecting articles submitted by influential employees of key advertisers. This author
knows of one instance in which a journal rejected an article by
a high-ranking official of an industrial firm and shortly thereafter
received a cancellation for a full-page advertisement that the firm
had been running for some time in the journal. In this case, however, advertising provided only a very small portion of the total
support for the journal. In the case of a journal that is wholly
dependent upon advertising, the publishers must certainly be
forced to give special attention to material submitted by employees of their major advertisers. Such a situation could easily
lead to the publication of articles slanted toward the products or
capabilities of those advertisers. Of course, there is nothing wrong
with a firm promoting its products and abilities, but the technical
paper is a particularly inappropriate medium for this purpose. Such
a practice would be unethical in the short run; in the long run it must
ultimately destroy the technical paper as a communication medium.

What can be done about this situation? Perhaps the question can be partially answered by looking at what should not be done. Certainly not much time and money should be invested in developing systems to improve the engineer's ease of access to information he cannot and will not use anyway. Instead of wasting energies on such a fruitless undertaking, we should attempt to close the gap between producers and consumers. The principal reason for the reluctance of engineers to use their professional society journals is very simple: they cannot understand the material published in them. Given the aim of bringing engineers into greater contact with their professional literature, there are two obvious approaches. The first approach, directed toward the engineer himself, is to enhance his mathematical sophistication so that he can understand the literature. The second approach is to bring the literature to the level that the engineer can use. The professional societies could publish a literature form whose technical content is high, but which is understandable by the audience to whom it is directed. This is not an easy task but it is not an impossible one. Some attempts have been made in this direction (for example, *IEEE Spectrum*), but they fall far short of solving the problem. Because these journals are generally designed to appeal to a very broad cross-section of a society's membership, the articles are either too broad themselves, or there are too few articles of a focused nature. The net result is that they are helpful but far from sufficient in bringing technical information to the average technologist. To attempt to change this situation, a large number of specialized journals, each with very intensive editorial attention, would have to be created. This is an expensive undertaking, but anything short of this scale simply begs the problem. The audience is certainly there, as attested by the success of the large number of controlled-circulation journals, which engineers do read. The task is not an impossible one. Engineers will read journals when these journals are written in a form and style that they can comprehend. Furthermore, technological information can be provided in this form. Why then do the professional societies continue to publish material that only a small minority of their membership can use? If this information can be provided in a form that the average engineer can understand, why haven't the professional societies done so?

The obvious answer to these questions is that the societies have only recently become aware of the problem. In the past, they were almost totally ignorant of even the composition of their membership, and they still know little of their information needs. Thus, they have never had the necessary information to formulate realistic goals or policy. Perhaps the most unfortunate circumstance that ever befell the engineering profession in the United States is that at the time when it first developed a self-awareness and began to form professional societies, it looked to the scientific societies, which had then existed for over 200 years, to determine their form and function. The decision to emulate the scientific societies must have been arrived at almost unconsciously. After all, most people still see little difference between engineers and scientists, and what better way is there to enhance the prestige of a new profession than to emphasize the similarities, not the differences, between it and an established high-prestige profession? So the professional engineering societies chose this path of least resistance and patterned themselves after the model of the scientific society. They have until this day ignored the differences between their membership and the membership of the scientific societies and have yet to properly define their role in society. The only reason the societies still exist is that most individual engineers can enhance their own prestige by regarding themselves as being "like scientists." Every engineer, from his freshman physics course on, regards himself in this way and covets the prestige given to the scientist. He therefore enrolls in a professional society, pays his dues, and dutifully lines his office walls with unread volumes of that society's publications.

Attempts to design improved information dissemination systems for professional engineering societies are very likely to lead to less than optimal solutions of the overall problem. These societies have failed to address the central question of the role of written communication in technology and the role of the engineering society in society at large. They have assumed that both these roles are well defined and accurately understood. This, however, is decidedly not the case, and until we have determined what they presently are and what they should become, we will have missed the essential prerequisites for discussion of such subsidiary problems as publication policy and information dissemination. Any steps to initiate a program of, say, editorial deposit will be premature without a thorough understanding of exactly what is to be accomplished through such an innovation, plus some estimates of the innovation's potential for accomplishing the organization's goals.

It is now high time that the engineering societies performed the task that has been awaiting them since their inception. They must discover who their member are and what their needs are. They must determine from this exactly what their role is. They must then decide what services they can offer their membership to aid in the fulfillment of this role. If it turns out that one of their goals is to provide for communication of certain specific kinds of information among their members, then they must search out mechanisms that will best accomplish that goal.

What would result from such a search for identity? First, many of the current policies of the professional engineering societies and their publications have no rational basis in and of themselves. They are mere copies of what the scientific societies have done in the past. An obvious policy of this sort is colleague refereeing of journal articles. Using this as an example, let us first examine the basis for it and then the implications that might be expected from its change or elimination.

Colleague refereeing is a practice that scientific journals developed in order to control their content. It is a filter through which scientific papers are passed; it is expected that only those of high quality and originality emerge. In practice, the refereeing process has taken on the additional function of certification of official recognition of a person's work by the scientific community.

In science, the colleague referees place their stamp of approval on the paper so that the author may be accorded his due recognition. Through refereeing, the scientific community formulates its decision of where to allocate the one resource it has under its control: peer recognition. This, of course, is why publication counts have become so important in determining promotion in the university system. This function is obviously ancillary to the communication goal of a scientific paper, but it can perhaps be justified on its own basis. If some other method could be found to perform the filtering function effectively, refereeing might be retained among scientists solely on the basis of the recognition function. This is much less true in engineering. There is nothing in engineering comparable to the social system in science. Most engineers gain recognition through means other than publication, and their need for certification is far less. Peer recognition is of only passing importance to the engineer who is concerned first and foremost with recognition by his organization and his advancement in it.

Referreeing is a quality control device. It is justified on the ground that it protects against publication of specious material and against the publication of unoriginal work. There is certainly a need to protect against publication of fallacious material. On the other hand, there is nothing wrong with publishing nonoriginal contributions in engineering. The first goal of engineering publication is to inform, not to stake out claims. False or poor quality material can be filtered at one stage by good editorial review coupled with quick-response outside review. Initial publication would be in the hands of both the professional societies and the controlled-circulation publishing houses. Follow-up reviews could then be compiled periodically by the professional societies that could fund and sponsor an engineering professor and one or more graduate students to review annually or biennially a particular technological field. The review would then filter out those contributions of poor or marginal quality that had slipped past the initial editorial review.

Under such a system the professional journals would move away from the scientific model and compete with the profit-making publications on a more equal basis. It would also make their material more timely since it would remove the long delay required for colleague refereeing. It is doubtful that any engineering journals will ever be able to publish material that is truly current with the state of the art. That the vast majority of engineers are employed by profit-seeking organizations, who justifiably or otherwise see early disclosure as a threat to their proprietary interests will mitigate very strongly against such a possibility. Engineers employed by university departments, on the other hand, while not deterred by such competitive forces, will probably not find it desirable to publish in nonrefereed journals. They, like their counterparts in science, are publishing not to communicate but to gain recognition. As long as engineering schools continue to use publication as a criterion of academic performance, their faculty will be motivated to seek out refereed journals in which to publish. Some will be forced to turn to journals of the scientific societies; others will be able to publish in those engineering journals that retain the old refereeing policy; still others will form their own journals or societies.

The net result will be that published material in engineering will never be abreast of the state of the art, and engineers will continue to rely overwhelmingly on personal contacts for current information. But the professional societies, by dropping refereeing, will have eliminated an unnecessary encumbrance, made their publications a bit more current, and enabled themselves to reach a far larger proportion of their clientele with relevant material.

Refereeing of articles is, of course, only one of many policies that might be changed as a result of a self-examination and role definition by the socieites. It is offered here only as an example, albeit an obvious one, of a policy that makes some sense for a scientific society but has little rational basis for engineers.

A question remains of just how the professional societies are to determine their role vis-à-vis their membership and society. In addition to data on the extent to which the professional society literature is used and the contribution of this literature to engineering performance (both of which are essentially negative findings), there are some data that show the relation between various forms of literature and the functions they serve.

Relating literature form to function is a first step in determining for the professional societies the implications inherent in changing the structure of their publications. Data from twelve technological projects (table 4.4) show that engineers use professional journals either for aid in the direct solution of a problem or when they feel a need to acquaint themselves with a new specialty or to broaden their competence. A rather surprising and interesting result appears in this table. When an engineer wishes to keep abreast of developments in his own field he refers not to his professional journal but to the trade journals; to keep abreast of developments on related or competing systems, he uses the privately sponsored engineering journals or *Scientific American*. One might have expected quite different results, say, professional journals for the engineer's field, privately sponsored journals for learning a new specialty, and trade journals for other systems. Certainly that would be the order of decreasing technical sophistication, and one would expect the engineer to seek reading material of decreasing technical content as he moved further from his own specialized area. One can infer from this that despite his technical sophistication, the engineer finds the professional journals in his own field unsuited to his needs. He is better able to obtain all of the needed information in his own field through personal contact. He uses the trade journals not so much to monitor the advance of his own technology as to determine the progress of competitors with new products, which he has learned of from other sources and which he is following to compare with his own situation.

Trade journals also supply information about new component

Table 4.4 Literature Channels Used by Engineers in Performing Nine
Functions

Function	Type of Publication Most Frequently Used
1. Aid in direct solution of a problem	Unpublished report, text, handbook, or professional journal
2. Determination of the results of related work performed by others	Unpublished report
3. Determination of procedures	Unpublished report
4. Learning new specialty, broadening areas of attention	Unpublished reports, professional journals, or *Scientific American*
5. Browsing that results in significant discovery	Unpublished report or trade journal
6. Verifying reliability of an answer	Unpublished report
7. Keeping abreast of developments in one's own field	Trade journals
8. Keeping abreast of developments on related or competing systems	Controlled-circulation engineering journals
9. Aid in definition of the operational environment	Unpublished reports

and subsystem developments that the technologist needs in his
work. That professional journals are used to some degree in
the direct solution of engineering problems and that *Scientific
American* is used to broaden the engineer's interests provide an
additional clue to what is happening. First of all, the engineer
does not use professional journals to keep abreast of developments in his own field. He learns of the general nature of such
developments through personal contact, with additional clues
from trade journals. Many of the trade journals have reasonably
competent industrial espionage systems and are able to uncover
information that firms are attempting to protect as proprietary.
The professional journals, obviously, are prevented access to such
information. Since state-of-the-art or review articles represent
a major part of the content of controlled-circulation journals,
these are used to provide information on systems that are related

to or compete with the engineer's own. They do not provide state-of-the-art information in the engineer's technical specialty so much as they provide knowledge of what is being done in competing product lines. When he wants to learn about another system or product line, the controlled-circulation journal can provide valuable information, which is considerably more detailed than that available in the trade journals. On the other hand, when he wishes to learn more about a new technical area, he does not use this type of journal. This is a bit surprising since these journals do contain some tutorial articles, but judging from the data, these are little used. Apparently the tutorial papers that have been appearing in increasing quantity lately in the professional journals better serve this purpose. In addition, *Scientific American* is also useful in this regard; it stresses the need for comprehensible presentations of complex material to aid the engineer in developing new skills. Perhaps the combination of articles from *Scientific American* to provide general concepts, followed by explorations in the more quantitive papers of the typical professional journal, is what is needed to best expand one's skill base in new directions. This is something that both the professional societies and the managements of technology-based companies should certainly look into in much greater detail.

Acquisiton of the Formal Literature

Before an engineer can use a written channel of information, he must somehow acquire access to it. The method by which acquisition is accomplished is quite easy to determine. For each literature reference in our survey, each engineer was asked to note the method by which he acquired it. Five possibilities were suggested, and a code was provided to the engineer to enable him to easily indicate the method he used.

From the results of this investigation (figure 4.6), it is quite clear that the majority of the engineer's formal reading material is maintained in the vicinity of his own desk. More than half of the formal publications acquired came from the engineer's

Figure 4.6 Method of Formal Literature Acquisition Employed by Engineers on Twelve Development Projects (Based on 134 Instances)

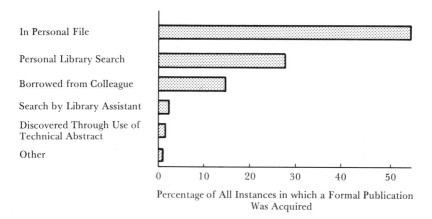

Percentage of All Instances in which a Formal Publication Was Acquired

own personal work area. If books and journals borrowed from colleagues are included, 92 percent of such acquisitions were accomplished informally. These also came from the immediate vicinity of the engineer's office or desk, having been borrowed from nearby colleagues. The engineers made precious little use of whatever bibliographic tools were available to them. Furthermore, this does not appear to be an exceptional situation. A number of studies have now shown a reluctance by engineers to make use of any library aids. There are two possible reasons for this situation. Engineers have either an improbable innate animus toward libraries or a background of library experience that has not been very rewarding. The latter possibility most likely stems from inadequacies on both sides of the checkout counter. Engineers seem for the most part to be ignorant concerning the use of bibliographic tools and to discount the potential of technical librarians for locating needed information. They seem to enjoy and prefer to tolerate the inefficiencies of searching themselves in order to do some browsing. There are very likely other good reasons for their reluctance to use the available bibliographic aids. Abstracts, for example, are not as well developed in most areas of technology as they are in chemistry, medicine, and psychology. Some engineers may have experienced unsatisfactory results in the few cases in which they have used technical librarians. The point, of course, is that we must examine both sides of the problem before we can formulate remedial action to bring the engineers and the library together on a satisfactory basis.

Use of the Company Library

The library is the second most important single source, but seldom does the engineer make complete use of its potential in his search. While 29 percent of the acquisitions in the study involved the library, in only 6 percent of the library cases was a library assistant used; the vast majority resulted from personal library search. Thus, the engineer's technique for acquiring formal literature is very informal. If the material is not available in his own work area, he either seeks it from a colleague or conducts a personal library search. Only on very rare occasions does he resort to such formal aids as a library assistant or technical abstract.

Although no explicit attempts were made during any of the studies to assess the merits of the libraries available to the engineers, the topic did come up several times during the interviews. In general, it can be said that as a potential information source, libraries were not a very salient consideration in the minds of any of the subjects interviewed. The typical response was, "Oh sure, we have a good library, but . . ." This was generally followed by an excuse for not using it, such as, "It's a pretty good walk over there," or by a contradiction: "They usually can't find what I want anyway." The value seen in using the library simply does not seem great enough to overcome the effort involved in either traveling to it or using it once the person is there.

In support of this point Frohman (1968) studied thirteen R&D groups in a large industrial firm and found that the extent of library use was an inverse function of the distance separating each group from the library. Furthermore, Frohman found very little variance among the thirteen groups. Individuals were also quite aware of the extent to which their fellow group members used the library, and there was very little variance within groups on the estimates of use by fellow members. This opens the possibility of group norms developing around library use. Physical distance to the library determines the mean level to which the group will use it. Members then adjust their behavior in order not to deviate too much from the group mean. Furthermore, there is fairly complete awareness and general agreement on the question of whether the group is a heavy user of the library. Physical distance is certainly a determinant of the extent to which an individual will use the library. But the development of a group norm, if this is true, will then make the effect of distance much more predictable. It becomes more difficult for the individual to deviate and so much easier to ride along at the accepted level. The subject of accessibility to an information source will be raised again in chapter 7.

Books versus Periodical Literature

Table 4.5 compares the mechanisms employed in acquiring books with those used for periodical literature and shows that books tend to be obtained through the library while journals are not. This finding would hardly be surprising if it did not contrast so sharply with the behavior of the scientists on the two research projects that were studied. The scientists relied on the library to provide them with journals (table 4.6) while most of the books they used they personally owned. At least part of the explanation for this must stem from the fact that a major portion of the journals the engineers used were of the free-subscription, controlled-circulation type. This removed the expenses of acquiring a personal copy. Such is not the case with scientific journals, which have high subscription costs. On the other hand, the scientist over his longer training period acquires a larger personal library than the engineer does, and in many companies, he finds it somewhat easier than the engineer does to have the company purchase new books for him, so he becomes a bit more independent of the library.

Table 4.5 Methods by Which Texts and Journals Were Acquired by Engineers (Twelve Development Projects)

	Number of Publications Acquired	
Methods of Acquisition	Texts and Handbooks	Journals and Abstracts
Formal		
Personal library search	8	21
Search by library assistant	3	0
Technical abstract	2	0
Total	13	21
Informal		
On desk or in personal file	8	65
Borrowed from colleague	2	18
Other	0	1
Total	10	84

Note: For total formal versus informal, $\chi^2 = 11.2$, $p < 0.01$.

Table 4.6 Methods by Which Texts and Journals Were Acquired by Scientists (Two Physical Research Projects)

Method of Acquisition	Number of Publications Acquired[a]	
	Texts and Handbooks	Journals and Reviews[b]
Formal		
Personal library search	10	18
Search by library assistant	0	0
Technical abstract	0	0
Total	10	18
Informal		
On desk or in personal file	36	16
Borrowed from colleague	6	5
Total	42	21

Note: For total formal versus informal, $\chi^2 = 7.58$, $p < 0.01$.

[a]Includes six double counts for books and two for journals. Double counts result when more than one method of acquisition was cited by a scientist.

[b]Reviews are included with journals as periodical publications since three of four citations were to *Reviews of Modern Physics*, a quarterly publication of the American Institute of Physics.

INFORMAL WRITTEN CHANNELS

Table 4.3 makes clear the importance of the unpublished report to the engineer. It is the principal written vehicle for transferring information in technology. Seldom does the information generated during the course of a major technological development reach the point of publication in any of the formal media.

This is not to say that nothing of this sort is published, however. A very large volume of written reports is normally generated. The extent of this documentation is reflected in the fact that in recent years, the cost of documentation has become a major component of the cost of large systems developments, and very often, even in the case of major aircraft developments, the cost of documentation outweighs the cost of the system itself.

Documentation takes on a wide variety of forms; at one extreme are scholarly publications recording advances in the state of the art and the solution of sophisticated analytic and mathematical problems, which could be published in the formal journal literature were it not for the restrictions of industrial or national security. At another extreme are documents of a very pedestrian sort containing, for example, detailed test procedures for system components or scheduling documents for the installation of equipment or the building of special facilities. For the most part the reports cited for the majority of the development projects in our present sample, which were preliminary design or feasibility studies, were generally more sophisticated in their content. [Unfortunately, we know little more than this about the contents; there is, however, something known about their origins (figure 4.7), the manner in which they were acquired, and the use to which they were put.] Rather surprisingly, documents that were generated within the organization were used to a slightly lesser extent than external reports. The division is nearly even, but one might expect in-house reports to be used much more than those generated by other organizations. As a result, there appears to be quite a flow of informal literature typical among organizations, at least those in the aerospace industry. Moreover, impressions gathered during the course of postproject interviews indicate that the customer agencies may be instrumental in maintaining this flow. This is not to say that the government laboratories are violating the proprietary interests of their contractors. It merely means that documentation produced by one company under government contract is made available to another firm working under different contracts. Normally the first firm might not be too concerned about the second firm's obtaining these reports but would not make any extraordinary effort to see that it did. The customer supplies a contractor with reports of his own and other government laboratories' work and with certain reports of other contractors. Indeed the customer appears to be responsible for providing nearly all of the reports in the "other company" category. The government laboratory, then, acts as an important force in promoting the flow of informal documentation among firms.

Figure 4.7 The Use of Informal Literature Sources by Engineers on Twelve
Development Projects (249 Instances of Use)

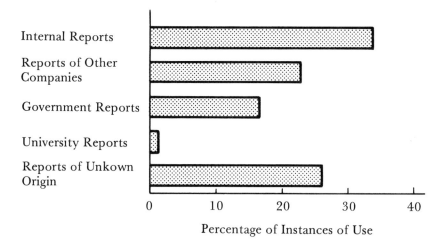

Percentage of Instances of Use

University reports were used hardly at all (figure 4.7). This may well result from the fact that little of the work done in engineering schools is of any direct relevance to industry. In addition, university reports suffer from many of the problems of the journal articles, which they eventually become. They are written for a very limited audience and cannot be understood by most engineers.[8]

The low use of university publications shows that engineers in industry rely very little upon current work performed in university departments or laboratories. Technology in the industrial sector is for the most part self-sufficient. It relies upon the university only to train its practitioners. The case in science, as one would suspect, is quite different.

Communication is poor in the other direction as well. The universities rely very little upon and are largely unaware of work performed in industry. A recent study of reading habits in one of the large MIT laboratories (Henize, 1968) finds that 61 percent of the unpublished reports read by a sample of sixteen engineers over a week's time (a total of twenty-two reports) originated within the laboratory itself. The remaining 38 percent (eight reports) were primarily government reports. While this is certainly a very limited sample from which to draw strong conclusions, it does not contradict our suspicion that university laboratories rely very little upon industrial technology.

Returning to the data presented in figure 4.7, the measure of relative use (the percentage of the total number of instances of use) correlated almost perfectly with the measure of the proportion of total reading time allocated. Figure 4.7 would be essentially the same regardless of which of these two measures was shown. The mean amount of time per instance once again shows very little variance among the different kinds of reports, ranging from 2.3 to 3.5 hours per instance. It is, however, significantly higher (mean of 3.2 man-hours per instance of use) than the mean duration of each use of the formal literature (1.4 man-hours per instance of use). This undoubtedly reflects the differing ways in which engineers use the two publication forms. They use formal literature more often for browsing and for looking up information on a relatively quick basis. They seldom browse through informal reports but normally read them very carefully in their entirety. Therefore, a longer period of time is required for each instance of use of the informal literature.

Functions of the Unpublished Report

Of the various kinds of unpublished reports, all but the documents of other companies are used principally in direct problem solving; the reports of other companies are used chiefly, and not surprisingly, to determine the results of related work done by others. University reports are also used to determine the results of work done by others and to assist in defining the operational environment in which a system must operate. The latter suggests that many of the university reports come not from the engineering school but from the science departments. They are used not so much to suggest solutions to technical problems as to better define the problem itself. From a problem-solving viewpoint, university reports are used primarily in such functions as criterion generation, criterion ordering (weighting), and the determination of acceptance limits for solutions.

Acquisition of the Informal Literature

Because unpublished reports occupy such a predominant position in the engineer's reading collection, it is interesting to know just how the engineer goes about acquiring them (figure 4.8). Once again, the informal acquisition mechanisms are more important than the library and its bibliographic aids. This time the finding is not as surprising. What is surprising is the importance that colleagues assume in supplying the engineer with reports. In considering the formal literature sources, it was seen that colleagues rank second and far behind the engineer's personal collection as a source of literature. With informal reports, however, colleagues assume the dominant role. The difference probably is attributable

Figure 4.8 Method of Informal Literature Acquisition Employed by
Engineers on Twelve Development Projects (Based on 147 Instances)

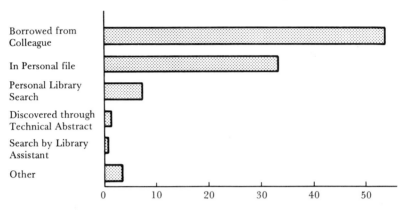

Percentage of All Instances in Which an Informal Publication Was Acquired

to the need for such reports to change hands many times over the course of a project. This is especially true of reports originating outside the laboratory. They are necessarily limited in number on the one hand and widely needed on the other, resulting in a situation in which they are passed back and forth among colleagues several times over the course of a project. Many engineers maintain a private library of reports, but their needs for information constantly vary. The turnover in private collections is quite high; the owner needs new reports as his work progresses, and colleagues borrow reports from the collection. The unpublished report seems to be in circulation constantly.

The Henize (1968) study, described previously, found that 69 percent of the seventy-one items of literature read by a sample of sixteen engineers over a week's time were obtained from colleagues.[9] This occurred either through normal routing of the document or when the recipient inadvertently saw it in a colleague's possession. Thus, there is a body of informal documentation that is in a state of constant flow within the laboratory. In this way, a single report very likely reaches a fairly large audience in a short period of time. All of this, one should remember, is within a single organization. The flow among organizations is a quite different matter.

There are very strong deterrents to the transfer of reports among organizations. First of all, considering those reports produced by industrial firms, there is the pressure of competition and the desire to protect proprietary interests. Over and above this, however, even when a report contains no information that the originator wishes to protect, there is very little motivation to promote transfer. The originator is unaware of the potential recipient's needs, does not really care about them, and is not motivated to find out or to try to meet the needs if he does find them out. Similarly, the potential recipient is usually unaware of the existence of the report and has no good mechanism for finding out about it.

A survey which we performed of thirty groups in U.S. government

laboratories working in fluidics technology indicates that half
of the individuals involved were unaware of the existence of 26
percent, or more, of the other groups. Moreover, fewer than half
had received reports from more than ten percent of the others.
In this case, there was little or no competitive pressure among
groups, but there was insufficient motivation to keep abreast of
activities in laboratories other than one's own. This situation
existed in fluidics despite the intensive activities of a governmental
interagency coordinating committee whose purpose was to promote awareness throughout the government laboratories. In technologies that lack this concerted effort at coordination, the situation must be at least as poor, and quite possibly it is far worse.

Origins of the Unpublished Report
Having seen the importance of the unpublished report in technology and having gained some idea of how an individual engineer
acquires and uses it, one might ask just how this particular device
came into being. Unlike the scientific paper, which developed
along with the norm of communality among scientists, requiring
them to share their research results with their colleagues, the
technological report developed in response to a quite opposing
set of forces: competitive pressures, the requirements of national
security, and the limited relevance of much of the information
generated as a by-product of technological activity. These forces
operated to require that the distribution of much of technology's
information output be given very limited distribution. This limited
distribution is accomplished through the control that can be
exercised over the dissemination of unpublished technical reports.

To a very large degree (even in this age of xerography), the distribution of unpublished reports can be controlled by the organization producing them. This is the reason behind their development. Because organizations for various reasons attach a condition of ownership to much of the technological information that
their employees generate and because they hope to sell this
technological information at some time, in one form or another,

they find it necessary to impose controls upon its dissemination. Such controls naturally preclude publication in the open literature. But because it is to the organization's benefit to have the information recorded for reference and disseminated within a limited domain (among employees, potential customers, and so forth), the vehicle of the organizationally sponsored informal report has arisen. These reports are issued in limited quantity for use within the organization and to selected individuals outside of the organization. In this way, the audience can be effectively controlled, at least to the extent of the first person receiving a report. Of course, there is no way of controlling dissemination beyond the first recipient. The originating organization cannot completely control the audience to whom this person passes the document or its contents, but by proper selection of first recipients the originator can pretty effectively circumscribe the audience that the information reaches.

Nevertheless, even within a limited audience this form of documentation performs a valuable service in propagating the technological state of the art. It is rather interesting to examine the way in which this process operates. Because of the availability of low-cost, high-speed reproduction machinery, an important document can be disseminated within an organization very rapidly. There would seem to be little control over the distribution within an organization. Once a document enters from the outside, it can be quickly reproduced and made available to anyone desiring it. The only point at which true control can be exercised is at the organizational boundaries. Once a document reaches another organization, however, the issuing body must rely upon individuals who are outside its immediate control to protect its interests. It becomes impossible to control dissemination *within* the recipient organization, but dissemination *to* other organizations is controlled, to some degree, by an interesting phenomenon: there seems to have developed in technology a norm against the transfer of another organization's reports beyond the limits of one's own organization. Very frequently the situation is encountered

in which an engineer possessing another agency's report will tell
someone from a third organization of its existence but with the
admonition, "You'll have to write to____ at ____ for a copy."
There is a very definite attempt to recognize another organization's proprietary interests. Of course, this must necessarily be
reciprocal. Violations of the norm are much easier to detect than
is the further dissemination of information within an organization.
When the originator learns that a third organization has knowledge
of the content of a report, the source of the leak can be very
obvious. This ability to detect violations is certainly necessary for
the development of the norm and explains the fact that it developed around the transmission of information among organizations rather than among individuals.

The importance of the unpublished report and some of these
peculiarities in its pattern of dissemination should clearly make it
a subject of great interest to communication research. There is
much to learn about this important vehicle, and research should
be directed toward determining its precise function and the roles
of various mediating agencies in its propagation.

The rather heavy interorganization flow of informal reports
must assume a large measure of the responsibility for maintaining
engineers' awareness of the state of the art. In every case, indica-
tions are that reports are used both in direct solution of problems
and in learning what others are doing. This, then, may be one of
the keys to the mysterious propagation of the state of the art in
technology and is worthy of a much deeper and more detailed
investigation.

COMPARISON OF FORMAL AND INFORMAL WRITTEN CHANNELS

The preceding discussion should demonstrate quite convincingly
the importance of the formal literature in technology. On the
average, engineers spend almost three times as much time with
such informal sources as industrial and government reports as they
do with formal literature sources. This finding varies, of course,

with the nature of the project; over the twelve projects in our sample, the ratio of time spent in using formal as opposed to informal literature ranges from 0 to 2.26. Comparing highly evaluated projects with their counterparts that received a low evaluation reveals no significant difference in the ratio of formal to informal publications used by either. In each case, high and low teams divide their time in about the same way between formal and informal literature. The nature of the technology appears to be the overriding determinant in the engineer's decision as to how he will divide his time between the two forms of literature.

NOTES

1. This was a compromise negotiated to obtain the cooperation of the two project managers.

2. The category "analytic design," which was originally intended to represent time spent in direct problem solving, was interpreted by most of the respondents to include group meetings and all activity other than information gathering that was of technical nonadministrative nature. In viewing these data, the reader should bear in mind that the proportions are of *time spent working on the project,* not time on the job.

3. Those engineers or scientists who produced technical solutions that were evaluated as being relatively high or low in quality.

4. The figure of 5 percent is significantly less than the 8 percent figures for all twelve projects. The reason for the differences is not clear but is most likely attributable to differences in the nature of the work. This, of course, illustrates the dangers of comparing different projects and clearly demonstrates the advantage of the matched pair or parallel project approach to controlling for problem solving.

5. The investigators began their research with this same popularly held assumption and naively asked each cooperating engineer to describe the way in which he conducted his literature search at the start of the project. Two years later, after having failed to find an engineer who would acknowledge having conducted an initial literature search, the question was dropped as a futile waste of time. The evidence, however, shows that despite their reluctance to call it a literature search, they did read considerably more at the beginning of a project.

6. *Literature* includes all written material: informal documentation, trade journals, books, and professional journals. The only written materials were some company documents, as explained.

7. On one project, this provision was added to the time allocation form at some point after work had begun. Literature references were obtained on

approximately the latter half of that project. The questionnaires by which these data were obtained are shown in the appendix to this book.

8. This is not to say that university research is worthless in the long run. That is a possibility that we can neither confirm nor deny with the present data. Nor is it to say that university research does not contribute to the long-run advancement of technology through the teaching function of the university. Certainly a great deal of good teaching is accomplished through research assistantships and through the incorporation of research results in classroom work. For students many of the most productive and interesting courses are those that incorporate some segment of current research work in their content.

9. Three of the seventy-one items were books; the remainder were informal publications of various sorts, including forty-four technical memorandums.

5 THE IMPORTANCE OF COMMUNICATION WITHIN THE LABORATORY

Most engineers are employed by bureaucratic organizations. Academic scientists are not. The engineer sees the organization as controller of the only reward system of any real importance to him and patterns his behavior accordingly. While the academic scientist finds his principal reference group and feels a high proportion of his influence from outside the organization, for the engineer, the exogenous forces simply do not exist. The organization in which he is employed controls his pay, his promotions, and, to a very great extent, his prestige in the community. He therefore behaves in ways that he feels the organization desires. This effect is demonstrated very clearly in his communication patterns, and particularly in the ways in which they differ from those of the academic scientist.

Much of what is to follow will be concerned with the effects of bureaucratic organization on engineers' behavior. This will involve an analysis of the costs and benefits associated with organizational communication and the problems confronted in moving new technology into the organization. The problems of keeping current with technological developments in the outside world is an acute one for bureaucratic organizations, which by their nature tend to be introverted. Even when the technology is available within the organization, many problems remain. It must be disseminated to the points where it can be used. There are many techniques and devices available for improving this internal communication. Among these are alternate organization structures, the use of informal organization, and architectural configuration of research laboratories.

ORGANIZATIONAL COLLEAGUES AS SOURCES OF INFORMATION

Conventional wisdom in research and development circles holds that there is a certain minimum size for an R&D laboratory. Many authors speak of a "critical mass," and some even contend that it lies in the neighborhood of about 1,000 employees. While this author has never seen any empirical evidence to support this contention (especially the magic number 1,000), it does make sense that there should be some minimum size below which a laboratory will fail to be self-supporting. The reason for this is quite plain. R&D today is a very complex activity. In most cases, the development of new products or processes requires a wide diversity of talents and knowledge. It is seldom that any single individual has all of the requisite knowledge. In the past several years, therefore, interdisciplinary project teams to deal with complex R&D problems have developed. Even such teams, however, seldom have all of the information needed to accomplish a project successfully. It is even rare when the diverse talents, experience, and technological understanding necessary to accomplish an R&D project can be found entirely within the small group of engineers assigned directly to the project. Few, if any, project teams can be entirely self-sufficient. Most R&D projects will therefore require some consulting support from people who are not assigned to them. At this point, the size of the laboratory becomes important. The one with a large and diversified technical staff should be better able to support its projects through internal consulting. In the laboratory without this resource, project engineers will have to turn elsewhere to satisfy their information needs.

THE TWIN PROJECTS

Data were obtained from the same projects described at the beginning of chapter 4. Again there were seventeen projects in eight matched sets and four projects for which there were no comparison partners. Measurements were made of the time spent consulting with organizational colleagues, the number and quality of ideas obtained in this way, and the degree to which internal consulting was used for problem-solving functions other than idea generation. In addition, with a second set of three pairs of parallel projects, the entire focus was on the internal consulting process and its relation to project performance. From these projects much more detailed data were obtained on both the extent and diversity of internal consulting.

CONSULTING WITH ORGANIZATIONAL COLLEAGUES

On the time allocation forms, each engineer was able to report time spent in communication with experts both inside and outside his own laboratory. Engineers divided their time almost evenly between the two. For the four pairs of projects on which relative performance evaluations were obtained, there appears some tendency for higher-performing teams to make greater use of internal consultants (table 5.1), but the difference between higher- and lower-performing teams is not statistically significant ($p > 0.10$). The higher-performing project teams on the average consulted more with organizational colleagues, but when performance on the entire project is the criterion, the difference in the amount of consulting is not sufficient to distinguish between high and low performers. Project performance is a very complex quantity, and a more valid measure of performance might be that of individual engineers on their constituent parts, or subsystems, of the project. Using this as the criterion produces slightly different results.

VARIATION OVER TIME

As in the case of literature use, high- and low-performing teams did not differ significantly in the total amount of time given to internal consulting. Unlike literature use, however, there were two peak periods during which internal consulting occurred. Literature was used very heavily at the beginning of a project, and then the use dropped off steadily toward project completion. Internal consulting followed a similar pattern at the outset, but about two-thirds through the project, it surged once again and reached its highest level (figure 5.1).

The reasons underlying the second peak in the consultation probably involve the need for help on certain types of problems that tend to arise at this point. Interface problems are very likely to occur at this point in a project. This is further supported by the fact that when changes in the probability levels reported on solution development records (an index of the extent to which

Table 5.1 Comparison of Time Allocated to Internal Consulting by Higher- and Lower-Rated Project Teams (Four Pairs of Parallel Development Projects)

Activity	Percent of Total Time Allocated		
	Total for Twelve Development Projects	Four Higher-Rated Projects	Four Lower-Rated Projects
Internal consulting	4.0	3.8	2.0
All communication (including internal consulting)	16.4	13.9	13.4
Total time reported (man-hours)	20,185	6,566	7,975

designs are being modified) are plotted over time, a similar pattern of two cycles of activity occurs. The second cycle of design changes correlates roughly with the internal consulting cycle. The cause of this midproject activity lies in the fact that following an initial period of coordination and interface tradeoffs, subsystems designs become relatively fixed and progress along lines that are largely independent of one another. As a result, minor changes are often introduced into subsystems without full knowledge of their impact on interfacing subsystems. As the project enters its final half, the need for coordination among subsystems designs once more becomes apparent. As designers of each subsystem learn that progress and changes in the design of adjacent subsystems have modified their interface requirements, they counter by modifying their own design and in the process seek out information upon which to base the new design. This effort very often requires not only coordination among the subsystems designers themselves but assistance from technical experts not assigned to the project team. That the second peak is more pronounced among lower-performing teams (figure 5.1) may reflect the fact that they allowed their interface problems to become more severe, or it may mean that their favored alternatives encountered difficulty and that they were searching (too late) for new approaches. Their internal consulting is more irregular than that of the higher performers, and between the two peaks, they

Figure 5.1 Comparison of the Proportion of Time Spent by Higher- and
Lower-Rated Project Teams in Consultation with the Laboratory's
Technical Staff

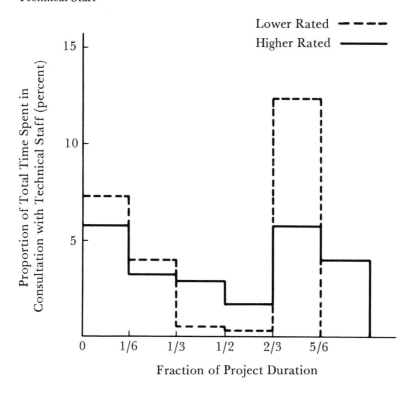

virtually cut off contact with colleagues outside of their project team. By staying in closer touch with organizational colleagues throughout the project, the higher performers may have obtained the necessary information to prevent such problems from getting too far out of control. The moral here is obvious: every project requires technical support beyond the project team. To be fully effective, such supporting contact must occur on a continuing basis and not simply when serious problems develop.

TECHNICAL IDEAS ORIGINATING WITHIN THE ORGANIZATION

There are two possible internal sources for ideas considered in the analysis. First is the individual experience of all members of the technical staff of the organization. Some of this derives from work performed while they were with the organization, but some also results from experience with other organizations and from technical training and education. The other major source of ideas lies in the information produced by the organization's other R&D projects, both current and past. Results or other data produced by some of these projects will, in many cases, suggest ideas for use in another project. This is somewhat different from the experience of the individuals in the organization and might be considered as something similar to organizational experience. A distinction is made between these two categories (figure 5.2), which were responsible for a nearly equal number of idea-generating messages.

Even when combined, the two information sources account for only a small proportion of the total number of ideas that were considered during the seventeen projects. All told, they were responsible for only eighty-one messages. This is only slightly better than the fifty-three emanating from the literature and is far short of the 300 provided by the customer and personal contact outside the organization. It amounts to less than 19 percent of all of the messages the engineers considered on the seventeen projects. The technical staff members of the laboratories engaged

Figure 5.2 Idea Sources for Solutions to Technological Problems (Seventeen R&D Projects)

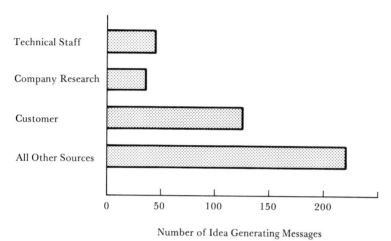

Number of Idea Generating Messages

in the seventeen projects were clearly underutilized as an information source. At this point it is good to bear in mind the size of the laboratories involved in this part of the study. The failure to consult internally cannot be attributed to any sparsity of technical staff in the laboratories studied. On the contrary, all are very large and self-sufficient. Most exceed the magic size of 1,000 professional employees, and all are housed in extremely large firms that at the time of the study were among the top fifty Department of Defense or National Aeronautics and Space Administration contractors. There must be some factor other than the availability of internal consultants militating against the use of internal consulting. This is a problem that will be considered later in the chapter.

INTERNAL CONSULTING FOR OTHER FUNCTIONS IN TECHNICAL PROBLEM SOLVING

Like literature, internal consulting can certainly provide for needs other than generating ideas. Internal consulting is relied upon quite heavily for the three functions of generating the criteria that define an acceptable solution, determining the limits of acceptance or rejection on each of those criteria, and finally testing possible solutions against the criteria (figure 5.3). In the projects under study, internal consulting was used quite heavily both to determine the criteria (better define the problem) and the levels of acceptance on the criteria. Personal contact within the organization provided nearly half of the messages for those two functions. Organizational colleagues were thus used more for problem definition than they were for idea generation. The technical staff of one's own organization can be particularly useful in this regard. These individuals should be able to interpret a technical problem in the context of the organization's business interests and in light of the prevailing technical culture in the organization. They understand the technical value system of the organization and the relative weight that this value system places on such criteria as cost, reliability, performance, and so on. Nevertheless, the surprising

Figure 5.3 Use of Internal Consulting in Three-Problem-Solving Functions (Four R&D Projects)

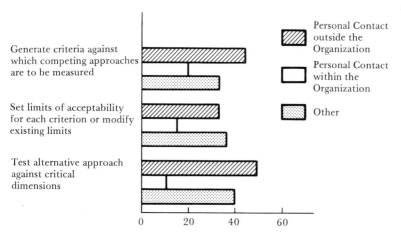

Proportion of All Messages Received For Each Function (percent)

fact remains that outsiders are called on more frequently than insiders to perform even these functions. So the problem remains. Personal contact within the laboratory is used less than personal contact outside, and this situation is especially pronounced in idea generation.

SOURCES OF IDEAS AND PERFORMANCE
The extensive use of personal contact outside the organization, while puzzling, might be explained if outside contact produced better information. In fact it does just the opposite.

When pairs of solutions to the same problem are compared in terms of their evaluated quality (figure 5.4), twice as many higher-rated solutions were based on consultation within the organization.[1] This is surprising in view of the fact that internal consulting was so little used. An attempt to explain this apparent paradox will be presented in some detail later. For now let us look further at the relation between internal consulting and the engineer's technical performance. The pattern exhibited in figure 5.4 looks reasonably impressive, but by itself it is not as convincing as the picture that emerges when a number of other bits of evidence are considered.

First of all, looking only at those solutions in the present study that were suggested entirely by organizational colleagues,[2] seven out of eight are among the higher-rated set of solutions. But in addition to the present study at least four others have independently discovered strong relations between performance and the use of personal contact within the organization. A study of twenty-two R&D proposal competitions (Allen, 1964) found a consistent positive correlation between proposal quality and the degree to which members of the proposal team had consulted with organizational colleagues who were not on the team. A study of sixty-four biological laboratories (Shilling and Bernard, 1964) found significant positive correlations between the degree to which informal discussion groups (an important mechanism for promoting internal consulting) existed in each laboratory and

Figure 5.4 Sources of Messages Resulting in Higher- and Lower-Rated
Solutions

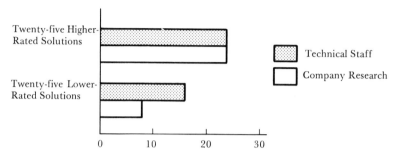

Percentage Based on Sources within the Laboratory

seven of their eight measures of laboratory "productivity and efficiency." Similarly, Baker et al. (1967) found that, despite a considerable amount of time and effort spent searching outside of their firm, "idea-generating groups" in a large electronics firm obtained their best ideas within the confines of the firm itself. Finally, Pelz and Andrews (1966) in their extensive study found that technical performance was strongly related to both the frequency and variety of an individual's contacts with organizational colleagues.

The results are all in clear agreement: internal communication is of overwhelming importance. On the average the best source of information for an R&D engineer is a colleague in his own organization. Several questions remain, however. What form should the consulting take? How important is diversity? Should the project members maintain contact with a wide variety of members of the laboratory's technical staff, or is it sufficient to restrict contact to a chosen few individuals in areas of greatest importance to the project? How important is communication among disciplines? And finally, what can be done to improve organizational communication?

THE INTERNAL CONSULTING STUDY

As an initial step toward answering some of these questions and furthering our understanding of the internal consulting process, a set of three pairs of parallel projects was chosen for intensive study (Gerstenfeld, 1967). Each of sixteen project members on the six projects reported weekly the extent of his contact with colleagues within his laboratory for the six-month duration of the projects. The reporting was accomplished through forms specially designed for this purpose (see the appendix to this book).

In order to understand better the nature and value of contact between members of the six project teams and their laboratory colleagues, the staff of each of the laboratories were divided into the following three categories:

1. Engineers and scientists in the laboratory who were assigned directly to the same project as the individual being studied.
2. Engineers and scientists in the laboratory who were *not* assigned directly to the project being considered but were members of the *same* functional group as the project member with whom communication was established.
3. Engineers and scientists in the laboratory who were neither assigned directly to the project being considered nor were members of the same functional group as the individual being studied.

Each of the categories will be considered separately in the analysis, and the extent to which project members consulted within each category was measured.

Level and Frequency of Communication

Level of communication was measured in terms of the number of times over the course of a project that each project member reported having communicated with a colleague on a technical or scientific subject. The communications could involve one or many colleagues; no consideration was made of the number of different people with whom an individual communicated. At this point we are more interested in determining the amount of contact between each project and the rest of the laboratory and determining whether the amount of contact bears any relation to project performance. In order to determine the relation with performance, the data from high- and low-performing subproblem pairs were compared as before. There are eight pairs of subproblems for comparison purposes.

Communication Within the Project

As an initial step, a comparison is made of the amount of communication within the project by high and low performers. Results show that high performers communicate far more with fellow project members (figure 5.5). The difference between highs and lows is statistically significant at the 0.05 level. Not too surprisingly,

Figure 5.5 Level of Communication Within the Project

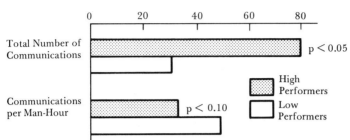

Communications per Man-Hour of Effort
with Other Project Members

though, the high performers also devoted a greater amount of time
to working on their subproblems (compare table 5.1). In other
words, they expended more effort than the low performers.
Naturally, if more time were spent on a problem one would question whether it is the level of communication among project
members which is leading to performance, or simply the greater
expenditure of effort. To answer this question, effort is held
constant in the lower bar of figure 5.5 and a comparison made
on the basis of frequency. With this done, the difference between
high and low performers disappears. The low performers actually
appear to have communicated a bit more frequently but the difference is no longer statistically significant.

As an individual spends more time working on a project, he
naturally communicates more with fellow project members. This
increased level of communication is but one of many activities
that increase with effort. Its contribution to performance, however,
cannot be separated from the contribution of these other activities.

Communication within the Laboratory
High performers showed a far greater reliance upon organizational colleagues not assigned to the project (figure 5.6). Once
again, the fact that high performers allocated more time to their
problems must be taken into account. This time, however, the
difference between the two remains strong even after the data are
normalized to account for the greater amount of effort allocated
by the high performers.[3] High performers not only communicated
more with their colleagues outside the project in absolute terms,
but their frequency of contact was greater as well. In both cases
the difference between highs and lows is highly significant statistically ($p < 0.002$). This result further strengthens the earlier conclusion of the importance of intraorganizational consultation to
technical performance. Project teams are absolutely dependent
upon technical information from beyond their membership,
and consultation with laboratory colleagues is the most effective
way of fulfilling this need.

Figure 5.6 Level of Communication with Technical Staff Not Assigned to the Project

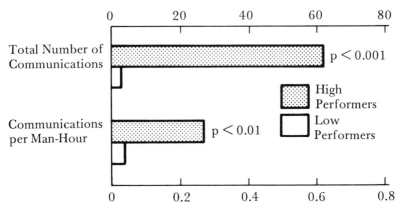

Communications per Man-Hour of Effort with
Laboratory Colleagues outside the Project

DIVERSITY IN COMMUNICATION

Now that we have seen that the remainder of the organization can contribute in important ways to project performance, we should explore the nature of this consulting process somewhat further to see just what kinds of support are most effective.

Colleagues outside a person's project may be divided on the basis of whether they are members of the same functional or disciplinary group (such as aerodynamics, structures, electrical power, chemistry) as the project member. When contacts reported by the six project teams are divided on this basis, it turns out that engineers producing higher-quality work communicated to about the same extent with colleagues both within and outside their disciplinary groups. Furthermore, they communicated with both to a significantly greater extent than did their less successful competitors. Engineers who turned in a poor performance seldom communicated beyond their own disciplinary groups. It is the higher-performing engineer who ranges out through the laboratory and extends his communication links far beyond his immediate work groups.

Sometimes the structure of the organization rather than the needs of the work determine communication patterns. For example, individuals may tend to communicate only with those with whom they are grouped organizationally, whether as a project team or as a functional department. Such people will not perform as well as their more adventurous colleagues who seek out necessary information far beyond the limits of their organizational groups. To be fully effective the engineer must be encouraged to develop contacts throughout his organization.

Another measure of diversity in contact is the number of different individuals with whom a project member maintains contact. The contribution of internal consulting to performance may be more or less dependent upon the *number* of colleagues with whom an individual communicates. Amount or frequency of communication, for example, may be unrelated to performance when that communication is directed to a single individual. Alternatively, the number of individuals with whom a project member communicates might even be inversely related to performance. The more people an individual talks with, the less time he has available to spend with each one, leading to a dilution of the total value of the information received.

Diversity Within the Project

Figure 5.7 shows that the average number of fellow project members with whom an individual communicates has no relation to performance. Over the course of their projects, both high and low performers consulted with an average of about two colleagues within the project team.[4] Once again, communication within the project seems to bear no relation to technical performance. Neither the number of fellow project members with whom an engineer communicated nor the extent to which he communicated with them shows any relation to the quality of the work on his individual subproblem. Project membership is a very strong determinant of communication, however. Members of a project will go to great lengths to maintain communication (Walsh and Baker, 1972) because their jobs require this effort. This being the case, it is possible that the projects that were studied had reached diminishing returns to intraproject communication. In contrast, there was very little requirement for extraproject communication. The projects were far from saturation in this regard. Project members could thereby improve their performance by increasing their exposure beyond the project.

It can be reasonably argued that increased communication within a project might contribute to better overall project performance without affecting the evaluations of any of the individual subsystems. It is subsystem quality, remember, that we have been measuring. Intraproject communication could, for example, reduce or prevent serious interface problems, which would contribute to overall project performance without being reflected in the individual subsystem evaluations that we received. The data, unfortunately, are incapable of testing this possibility. Certainly the elimination of all communication among project

Figure 5.7 Diversity of Communication Within the Project (Eight Subproblem Pairs)

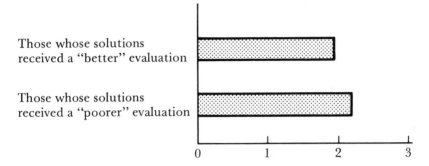

Those whose solutions
received a "better" evaluation

Those whose solutions
received a "poorer" evaluation

0   1   2   3

Mean Number of Fellow Project Members
with Whom Communications Were Reported

members would be disastrous. So communication contributes to performance at least up to some level. In the present situation that level may have been reached in both high- and low-performing projects.

Diversity Within the Functional Group

Project members producing better technical solutions communicated with more colleagues who, although they were not assigned to the same project, were in the same group as the project (figure 5.8). This difference is highly significant statistically ($p < 0.001$).

The importance of contact with colleagues in the same specialty presents an argument for functional forms of laboratory organization in which project members remain both physically and organizationally attached to their functional department. This, as we will see, permits easier communication among project members and their disciplinary colleagues. In related research, Marquis and Straight (1965) found that of thirty-eight major R&D projects, those organized along functional lines attained better technical performance than those organized on a project basis. The present data complement Marquis and Straight findings. When located physically and organizationally with their functional colleagues, project members will be more inclined to communicate with them, and this communication can be highly benefical to the performance of the project. Later in the present chapter, I shall discuss the various tradeoffs that are possible among different organizational forms.

Diversity of Contact Within Other Functional Specialties

Only one of the engineers whose solution was poorly evaluated reported any contact with a colleague who was neither a project member nor a member of his functional group. Moreover, this individual reported communicating with only one person. In contrast to the low performers, those submitting highly evaluated solutions (figure 5.9) reported an average of twenty times more

Figure 5.8 Diversity of Communication Outside the Project But Within
the Functional (Disciplinary) Group (Eight Subproblem Pairs)

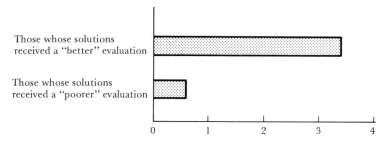

Mean Number of Disciplinary Colleagues outside the
Project with Whom Communications Were Reported

Figure 5.9 Diversity of Communication Beyond Both Project and Functional
Group (Eight Subproblem Pairs)

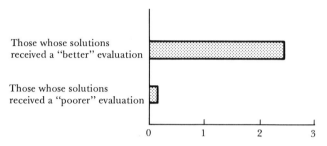

Mean Number of Colleagues Who Are neither
Members of the Project nor of the Same
Functional Group as the Project Member

contacts with people outside their specialty. Equally important is the observation that, in general, project members contacted few people in functional specialties other than their own during the course of a project. On the average, over the duration of the six projects, each engineer contacted fewer than two individuals on the laboratory technical staff who were not members of either his own project or functional group.

INTENSITY VERSUS DIVERSITY

Since both the level of communication activity and the number of people outside the project with whom an individual communicates are strongly related to performance, the question remains whether one is the principal contributor. There is, of course, the possibility that both contribute equally. For this reason Kendall Tau rank-order correlations were performed on the data to allow us to examine the relationship between two of the three variables while holding the third constant.

In this case three variables were intercorrelated: number of communications with colleagues outside the project; number of colleagues with whom the individual communicated (outside the project); and performance on the subproblem. The partial correlation between performance and number of communications, holding number of colleagues outside the project constant, and the partial correlation between performance and number of colleagues, holding number of communications constant, are compared in table 5.2. One might guess from this table that frequency and diversity of communication are themselves highly correlated, and, in fact, they are (Tau = 0.86). Project members did not increase the scope or diversity of their communications at the expense of depth or amount of communication with each individual, so there is no dilution effect as a result of broadening one's communication contacts. In fact there is some slight indication that the opposite occurred. The number of communications per person increased somewhat as the number of people with whom an individual communicated increased. The nine individuals

Table 5.2 Relation of Two Measures of Communication Outside the Project to Subproblem Performance

Correlation between Solution Quality and:	Kendall[a] Tau	Kendall Tau (Partialled)	Variable Held Constant
Number of communications with colleagues outside the project	0.75[b]		
		0.31	
			Number of Colleagues
Number of colleagues communicated with outside the project	0.75		
		0.31	
			Number of Communications

[a]The Kendall Tau is a rank-order correlation that measures the degree of association or correlation between two variables. It ranges from –1.0 (a perfect inverse relationship between two variables) through zero (no relationship) to +1.0 (perfect direct relationship).

[b]$p < 0.01$.

who reported communications with only one or two persons outside their project had an average of slightly under five communications per week with each of them. The level of communication increased as the project member's contracts became more diverse, and both were related to performance on the project

SUMMARY OF THE EVIDENCE
Despite the hopes of brainstorming enthusiasts and other proponents of group approaches to problem solving, the level of interaction within the project groups shows no relation to problem-solving performance. The data to this point lend overwhelming support to the contention that improved communication among groups within the laboratory will increase R&D effectiveness. Increased communication between R&D projects and other elements of the laboratory staff were in every case strongly related to project performance. Moreover, it appears that interaction outside

the project is most important. On complex projects, the inner
team cannot sustain itself and work effectively without constantly
importing new information from the outside world.

In the beginning of the chapter we saw that such information
is best obtained from colleagues within the organization. In addition, high performers consulted with anywhere from two to nine
organizational colleagues, whereas low performers contacted
one or two colleagues at most. This suggests that increasing the
number of colleagues with whom an engineer consults contributes independently to performance. In the present situation, the
number of colleagues contacted correlates so strongly with the
number of contacts made that it is extremely difficult to separate
the two for analysis. Given this limitation, figure 5.10 best sum-
marizes the inferences that can be drawn from the data.

As the number of consultants increases, so do the number of consultations, and the two operate in series to produce higher project
performance. Concurrently increasing the number of consultants
contributes to performance somewhat independently of the number of consultations. These findings are in general accord with
those of Pelz and Andrews (1966), who concluded from their data
that

Frequent contacts with many colleagues seemed more beneficial
than frequent contact with just a few colleagues. Similarly, having
many colleagues both inside and outside of one's own group
seemed better than having many colleagues in one place and just
a few in the other. So anything you can do to promote these
forms of contact should be in the right direction.

Further support for the Pelz and Andrews' diversity hypothesis
can be found in the present case. All of the eight high performers
reported consulting both with members of their functional group
and with colleagues from other functional specialties. The remaining two restricted their consultation within the functional
group. None of the remaining seven consulted with individuals
whose functional specialty differed from their own.

This closer look at three pairs of parallel projects provides

Figure 5.10 Contributions to Performance of Breadth and Intensity of
Internal Consulting

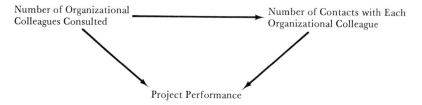

further confirmation of the value of promoting personal contact among project members and other staff of the laboratory. I will return again to a consideration of internal consulting in the final chapters, but first I will turn to another possible source of technical information, personal contact with parties outside the organization

NOTES

1. The difference between proportions is statistically significant ($p < 0.01$).

2. Remember that ideas are developed through a convergence of messages from many sources. Some sources for the same idea may be both within and outside the organization.

3. The term *high performer* is used for convenience in presentation. The reader should be reminded that the evaluation was made of a person's work on a specific job and not of the individual himself. Some of those labeled "high performers" may very well have not been regarded as such by their organizations. They merely did a better job than their counterpart on this one occasion. The converse may be true of course for "low performers." This subject will be returned to when we take up the problem of causality.

4. Teams were comparatively small and averaged about six members.

6 COMMUNICATIONS AMONG ORGANIZATIONS

An organization faces two problems with respect to scientific and technical information: acquisition and dissemination. Chapter 5 showed the importance of proper dissemination within the organization. This chapter will confront the acquisition problem and try to determine the means by which information is best transferred among organizations.

Although the evidence so far has emphasized the importance of internal communication, no R&D laboratory can be entirely self-sufficient. In recent years, a number of industrial and government R&D establishments have reached mammoth proportions, employing thousands of professionals. But not even these extremely large organizations can long sustain themselves without replenishing their store of knowledge from outside. In order to develop in its cumulative fashion, technology depends upon the blending together of contributions from technologists throughout the world. It is inconceivable that any single laboratory, in even a narrowly defined specialty, could or would even want to replicate all that goes on in the world around it. The laboratory operates as an open system. It must import information from its environment in order to sustain life. All laboratories are therefore dependent upon other laboratories. This holds true even in a competitive industrial climate. In the present chapter, I will explore the ways in which laboratories accomplish this life-sustaining importation of information.

OUTSIDERS AS SOURCES OF INFORMATION
There is great variety of the types of personal contact that an engineer can develop outside his laboratory. In our studies, the outside sources ranged from experienced consultants who had worked with an organization for many years to the sales representatives (vendors) of other companies who would like to sell components, subsystems, or instrumentation to the project. Apart from the customer organization that sponsors the project, the greatest proportion of messages from outside the laboratory come from vendors. This is to be expected. Vendors usually do not

have to be sought after; they will take the initiative of approaching the project engineers. Next to vendors in the amount of information supplied are what might be called unpaid outside consultants, usually members of government, nonprofit, or university laboratories, who are consulted on a brief and informal basis. The Jet Propulsion Laboratory of the California Institute of Technology is a good example of a laboratory whose members are a very rich source of ideas for the aerospace industry. Project engineers confronted with a problem will often visit one of these laboratories for general discussions with engineers who have worked on similar problems. The reason they choose government, nonprofit, or university laboratories is, very simply, that these are more readily accessible than other industrial laboratories. Competitive pressures usually foreclose the possibility of an engineer from General Electric visiting Westinghouse to discuss the common problems of what very well may be competing products. The nonindustrial laboratories do not feel the pressure of this sort of competition, or at least believe that they should not feel it, so they are more willing to sit down and discuss problems and solutions.

A third category of outside source is the paid consultant. These were relatively rare in the study and were usually university professors called upon for their knowledge in a highly specialized area. The nature of the relationship with the paid consultant was quite varied. A few had been used by the laboratory for many years to provide a unique competence that the laboratory had been unable to obtain through normal recruiting. Others were engaged only to provide competence needed on a specific project. Usually the consultant was used not to provide solutions to problems but rather to assist in defining the problem itself.

THE TWIN PROJECTS

Just as in the previous two chapters, data were gathered from seventeen R&D projects that formed eight matched sets and from an additional four unmatched projects. The data show the proportion of time spent with outside sources, the number and quality of ideas received from them, and the use of outside contacts for other problem-solving functions.

TIME SPENT IN PERSONAL CONTACT OUTSIDE THE LABORATORY

Project engineers allotted about 5 percent of their total time and about 33 percent of their communication time to outside contact (table 6.1). The difference between high- and low-rated performers in the extent of their contact outside the laboratory is not statistically significant. Looking at the way in which this time was distributed over the duration of the project again shows some difference between high and low performers and a marked cyclical effect (figure 6.1). The poorer performers used outsiders heavily both at the outset and again just beyond the mid-point. Just why they spent so much time with outsiders at project initiation is unclear. It does appear, however, that the higher performers were once again more consistent in their contact with the information source. The poorer performers either spent a relatively high proportion of their time with outsiders or none at all. The higher performers spent about the same proportion of their time regardless of what point they were at in the project. This was true also in the case of internal consulting, and to a lesser extent with

Table 6.1 Time Allocated to Consulting with Sources Outside the Laboratory

	Percent of Total Time Allocated		
	Total for Twelve Development Projects	Four Higher-Rated Projects	Four Lower-Rated Projects
Consultation outside the laboratory	4.45	5.09	5.32
All communication (including consultation outside the laboratory)	16.40	13.90	13.40
Total time reported (man-hours)	20,185	6,566	7,975

Figure 6.1 Comparison of the Proportion of Time Spent by Higher- and Lower-Rated Project Teams in Consultation Outside the Laboratory

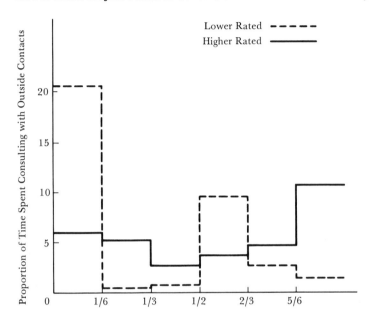

literature. Thus there appears some benefit to maintaining consistent, continuous relations with information sources of all types.

IDEAS FROM OUTSIDE SOURCES

A greater number of ideas for solution were suggested by sources outside the laboratory than by all other sources combined (figure 6.2). This is true even when ideas from the customer are excluded. If messages from the customer were included in the count of messages from outside, then these sources were responsible for two-thirds of all idea-generating messages received by the project engineers. This demonstrates convincingly the degree to which R&D projects are dependent upon outsiders. It might be argued that the reason for turning to outside sources lies in the inability of the laboratory's technical staff to provide the needed information. In part this is tue, but it is an incomplete explanation. First, there is nothing peculiar about the organizations in the study that would make them more inclined than the average to resort to outside sources. On the contrary, the laboratories in the study are among the very largest and most highly regarded in the United States. If the technical staff of these laboratories were incapable of supporting the information needs of projects, then the situation must be far worse in the average R&D laboratory. Instead, some other factor must be at work here. Very often even when the required competence should have been available in house, a project engineer still turned first to an outside source. His reasons were many. Often he was not aware of the inside contact in time. Other times he would rationalize going to the outside source on the basis of its particular experience being uniquely appropriate to his need. I strongly suspect that what is happening here is simply another manifestation of the effect of differences in the cost of using different information channels, which was discussed in some detail in chapter 5. An outside source for many reasons is often more accessible and less costly to use than the internal contact. The same psychological cost does not have to be paid. This is especially true of vendors. Many engineers approach

Figure 6.2 Idea Sources for Solutions to Technological Problems
(Seventeen R&D Projects)

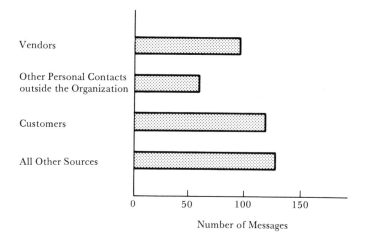

Number of Messages

design problems by simply calling upon their appropriate vendors, passing most of the problems off on them, and then accepting the best solutions. Vendors are quite willing to cooperate in the hope of making a sale, but it is doubtful that any very innovative solutions come of this tactic. Vendors are perhaps the easiest of outside sources to contact; even those with nothing to sell are often more easily approached than the colleagues within one's own organization. For many reasons the outsider is often more approachable and more willing to give of his time than is the organizational colleague. As a result, the typical engineer is quite prepared to search outside his organization when in need of information.

USE OF OUTSIDERS TO AID IN PROBLEM DEFINITION
The three functions depicted in figure 6.3 are all related in one way or another to problem definition. The first function, criterion generation, specifies the dimensions, cost, power output, speed, weight, and so on, which are of key importance in the problem. It defines the problem in a multidimensional space. In performing the second function, the engineer decides just what levels of performance on each dimension will be required of an acceptable solution. In the third function, each alternative is tested against the set of criterial dimensions and limits. It was not so surprising in chapter 5 to see organizational colleagues so heavily involved in all three. It was a bit surprising to see the extent to which vendors and others from outside the organization are involved in these processes. This can be partially explained by the fact that vendors must often be consulted to determine the performance limits of the hardware they would provide. Sometimes a number of vendors are asked to provide the performance limits of their equipment, and from these data plus some quantity of engineering judgment, the project engineer decides upon the level of performance that his solution must provide. Vendors thereby determine many of the constraints on the engineer's problem. Others from outside are often consulted from their experience and understanding

Figure 6.3 Use of External Contacts in Three Problem-Solving Functions (Four R&D Projects)

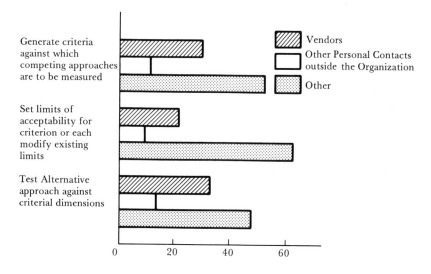

of similar problems and in this way contribute to problem definition.

In addition, both paid and unpaid outside consultants are often called upon to specify the characteristics of the operational environment—for example, the atmospheric and soil conditions that will be encountered in certain locales—and these in turn specify dimensions and limits for the system or subsystem. These consultants have much to offer during this stage of the design process. What is suprising is the overall extent to which vendors are used (figure 6.3). Engineers are apparently quite inclined to make full use of any free service proffered them.

THE CONTRIBUTION OF OUTSIDERS

Once again we have the opportunity to evaluate the contribution of a class of information channels by performing a comparison between high- and low-performing teams (figure 6.4). High performers in this case obtained far fewer of their idea-generating messages from outside the laboratory organization. Furthermore—and this is why vendors and outside consultants are shown separately—the difference is solely in the use of consultants.[1] Low-performing teams relied far more than high performers on personal contacts other than vendors. When one considers only those solutions that were suggested entirely by outsiders (either consultants or vendors)—and there are fourteen of these—only five are among the higher-rated set.

These results are surprising in two ways. First, there is an apparent inverse relationship between performance and the use of information sources outside the organization. Second, of all the outside sources, vendors might be expected to provide the least reliable information; yet it is not the vendor but other forms of personal contact that are responsible for the inverse relation. I will attempt to explain each of these observations in its turn, but first let us see whether this is an isolated, and possibly unique, finding or whether there is additional evidence to show the poor performance of external contacts.

Figure 6.4 Solution Quality as a Function of Information Source (External Personal Contact)

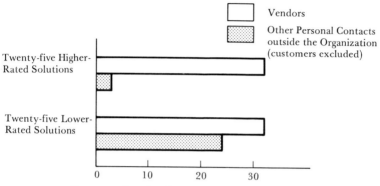

Percentage Based on Sources outside of the Laboratory

There are four studies in addition to the present one that relate individual or organizational performance to the use of personal contacts outside the organization. Baker et al. (1967) looked at a number of idea-generating groups in a large electronics firm. The groups were charged with the responsibility of generating ideas for new products and were provided with substantial travel budgets to seek out such ideas widely. Upon completion of the survey, it was discovered that ideas that had originated within the firm far surpassed in quality those coming from outside

Shilling and Bernard (1964) in a study of sixty-four industrial and governmental laboratories in the biological sciences found a strong inverse relation between the extent to which the laboratories used paid outside consultants and laboratory performance. They had eight measures of laboratory performance and found statistically significant negative correlations with all eight measures.

In an earlier study (Allen, 1964) I found similar results. In a study of twenty-two competitions for government R&D contracts, I found an inverse relation between the extent to which proposal team members consulted with persons outside of their respective firms and the technical quality of the proposals they submitted.

In the one exception to this general trend, Hagstrom (1965) in a study of 179 prominent researchers in the formal (mathematics, statistics, and logic), physical, and biological sciences found a strong positive correlation ($Q = 0.85$) between communication outside the individual's university and productivity in terms of papers published. The correlation between productivity and communication within the university department was only about half as strong ($Q = 0.42$). Although this appears to conflict with the other findings cited, I believe that when certain conditioning factors are taken into account, it is really concordant with them. In Hagstrom's case communication outside the organization was restricted to the field of the communicator. Hagstrom measured only communication with others in the same field,

perhaps the same "invisible college." It is probably safe to say
that for university scientists this is a far more salient social system
than is the university that employs them. To a very great extent
it controls the reward system of the scientist. It does this through
the allocation of grants by the panel or referring process, the
control it exerts over whose papers are published in which journals,
and even to some extent over the promotion process within the
university. So it is to this social system rather than his employer
that the university scientist accords his loyalty. Since this social
system crosses the organizational boundary, it reduces any effect
that organization has on extramural communication.

A question remains of why the organization should interfere
with outside communication in the first place. Here we must
return to one of the essential differences between industrial technologists and university scientists that was pointed out in chapter
3. In the first four studies in which external information sources
performed poorly, the organizations were all more or less bureaucratic in their structure—that is, they were hierarchically organized
and had clear division of labor, work procedures, differential
rewards by position, and most of the other attributes proposed
by Weber and others to define the bureaucratic model. A university faculty does not fit this model at all; hence, the exception in
the case of Hagstrom's data.

In addition to the above characteristics, a bureaucracy demands
an employee's loyalty to an extent that is not even approached in
a university. This (and its proprietary interests resulting from a
competitive environment) causes it to restrict the flow of information across its boundary. It follows that the differences in the
effectiveness of extraorganizational communication in the two
situations can be attributed in large part to two factors: (1) the
relative commitment of the individual to the organizations or
social systems at hand and (2) the degree to which the boundaries
of these organizations are formally structured.[2]

Hagstrom's scientists had little difficulty communicating
across the bounds of their academic departments so long as they

remained within their own disciplines. The importance of the academic department to the university scientist in no way compares with the importance of the employing organization to the engineer working in industry. Therefore, Hagstrom's scientists confronted a low impedance in communicating across the bounds of their academic departments because the academic department elicits a lower degree of commitment from most academic scientists than does their professional discipline, or "invisible college." Had Hagstrom examined communication across disciplines, he might have found a higher impedance than at the bound of the academic department but not so high as at the periphery of a more structured organization, such as an industrial or government laboratory.

Key among the problems encountered by the bureaucratic form of organization is its tendency to isolate itself from the outside world and to erect barriers to communication with the outside world. This results from the need on the part of the organization to partially isolate itself from the rest of society and to exercise control over those situations in which interaction with the outside world occurs. In addition, and perhaps more important for the present purposes, every social system develops over time a common way of viewing the world and a common interpretation of the tasks that it faces. Competitors in a single industry often face identical problems and yet produce markedly dissimilar, sometimes even predictably dissimilar, solutions to those problems. The differing cultures of the organizations interpret or structure the problems in different ways; they weigh the solution criteria differently and thus almost guarantee the development of different solutions. Some laboratories, for example, are noted for the conservatism of their designs; others are gamblers and are noted for creative thinking and occasional outstanding breakthroughs. These characteristics, as well as certain of the long-run organizational goals, become ingrained in the members of the organization and thereby provide a conceptual or coding scheme by means of which they categorize their world and communicate

about it to others. This common coding scheme develops in part
as a result of the organizational members having experienced over
time a series of common experiences and in part by a philosophy
developed out of the organization's early experiences or resulting
from the influence of a few key individuals. The coding scheme is
a common viewpoint manifested in a shared language and a common set of attitudes. It enhances the efficiency of communication
among those who hold it in common but can detract from the
efficiency of communicating with anyone who follows a different
coding scheme. Engineers in an organization are able to communicate better with their organizational colleagues than with
outsiders because there is shared knowledge on both ends of the
transaction and less chance for misinterpretation. It is amazing to
see how misinterpretation can creep into the communications
between organizations. This is what Bar-Hillel and Carnap (1953)
call "semantic noise," that is, the impedance that causes error in
the interpretation of messages and is analogous to noise sources
in physical systems that cause error in reception of messages
(Cherry, 1957).

Anyone can testify to this who has participated in an inter-
organizational meeting or negotiation and watched the participants
talk right past one another for hours or days without making any
progress simply because they were "not speaking the same language." Different organizations view the world differently and so
do their members. This creates an inherent problem whenever
communications must take place across an organizational boundary.
Of course, the coding schemes are far less exclusive than most
languages. There is a great deal of overlap among the coding
schemes of different organizations operating within the same
culture. On the other hand, the nonoverlapping areas, however
small, can potentially operate to produce semantic noise, and they
can be even more troublesome because it can go undetected.
Anyone who does not speak French knows his deficiency, but
very often we think we know what someone from another organization is saying when in fact our understanding is very different

from his. Organizational coding schemes both enhance the efficiency of communication among those who hold them in common, and detract from the efficiency of communication between holders of different coding schemes. As Cherry (1957) tells us, "The semantic-information content of a statement (which includes all that is logically implicit in that statement) is available only insofar as the rules of the language system are known." In other words, communication between coding systems, without knowledge on the part of one or both communicators of the other's coding system, introduces the possibility that part of the semantic-information content of the message will be lost.

At this point it might be helpful to review once again some of the empirical results that led to this point. First, several studies of industrial and government scientists and engineers revealed an inverse relation between extraorganizational communication and performance. This contrasted with a direct relation between intraorganizational communication and performance. Second, in Hagstrom's study, where the organization (an academic department) appeared to occupy a subsidiary position to a more inclusive social system ("invisible college" or academic discipline) and where the communication process measured was external to the first entity but internal to the second, a strong positive relation was found between the extent of communication and performance. Third, the negative correlations between external communication and performance cannot be taken to imply a causal relation. From a purely logical viewpoint, it is probably not information but lack of it that leads to poor performance. The engineer seeks information because he perceives a potential gap in his knowledge relative to a specific problem. If this gap in fact exists, his performance should be better if it is filled; it should be worse if it is not. Some information sources are consistently better than others at filling these gaps. According to the evidence presented, for those in bureaucratic organizations an organizational colleague is generally more likely than someone outside the organization to be able to fill this need. This is due less to a lack of competence by the outsiders than to a problem in communication with them. The difficulties encountered in communicating across organizational boundaries are especially critical when the relation is of short duration. Over a long time period, people operating in two coding systems can become aware of the precise nature of the difference between the systems. They can then translate information from one into the other and are less bothered by the semantic noise problem. At least in the projects reported in figure 6.4, the relations with outsiders were of very brief duration. This is also generally true of the proposal competition study and probably of the study by Baker et al. (1967). There is no way of telling whether the paid consultants in Shilling and Bernard's study were engaged on a long-term retainer basis or brought in for specific problems. At any rate three of the four studies probably analyzed the use of outsiders on a short-term basis where they were used to provide answers to specific problems. In view of the communication difficulties just discussed, this is probably the worst possible arrangement. If outsiders are needed to supplement the staff of the organization in particular areas, this support should be planned and engaged on a long-term basis. In this way they can be socialized to some degree into the organization and learn the nature of its business, the value system within which it functions, and the coding system by which it communicates. Over time the outsider becomes indistinguishable from any of the regular organization.

THE TECHNOLOGICAL GATEKEEPER

Even though consultants can eventually be indoctrinated in the culture of an organization, the use of consultants cannot satisfy all of a large organization's needs for communication with the outside world. Even the largest R&D laboratories support only a very minor portion of the world's scientific and technological activity. They must, somehow, keep their personnel aware of the results of research done elsewhere. The question remains: how?

Acquisition of technical documentation and literature certainly does not appear a promising solution if no one is going to read it. Contact with most outsiders does not appear to be any more effective, albeit for different reasons. Yet a laboratory must some- how accomplish this transaction with its environment if it is to survive. As I have argued in the last few paragraphs, long-term consultants provide some help. Normal turnover of personnel also helps to transfer information because the new employee often brings new or different technology with him. These mechanisms, however, do not seem to be sufficient to meet fully a technical organization's needs for outside information. In fact, the lab- oratory studies were first initiated to explore this problem.

In the laboratory studies, a measurement was made of the structure of an R&D laboratory's communication network, or network of personal contacts. At first, this was done by simply asking people to list the names of those people in their organiza- tion with whom they discussed technical or scientific matters most frequently A base frequency of at least once per week was specified. Later a sampling method was used. Under the latter system, individuals were asked once a week, on randomly chosen days, to check off the names of those with whom they had tech- nical or scientific discussions *on that day*. This was done for a period of several months, and networks were then based on average communication frequencies of once per week or more often. The choice of a once-a-week frequency is purely arbitrary, although it does seem to represent a fairly heavy degree of con- sistent communication. Plots could be made of any other fre- quency of communication just as easily.

The two methods of determining communication patterns produce comparable results. Some comparisons have been made using both approaches in the same organization. The only im- portant difference seems to be that the time-sampling approach produces a more highly connected network. When people are asked to list those with whom they communicate most frequently, they very seldom list anyone who would not have appeared had the time-sampling approach been used. They do forget some of the people with whom they communicate regularly, though. Perhaps this reflects a difference in the importance of some communication links. The one-time listing of names may filter out some of the less important communication partners that the time-sampling approach includes. Because the results from both approaches are fairly comparable, data from both methods will be combined in the analysis that follows.

A typical communication network of a small (thirty-six professionals) R&D organization is shown in figure 6.5. Each of the professionals in this organization is represented by a circled number. Arrows are used to indicate which individual reported communications. For example, number 5 reported communication with number 4 at a frequency of at least once per week. Number 4 also reported at least that frequency of communication with 5. Arrows consequently are shown in both directions. This is not always the case. Number 15, for instance, reported communication with 17 more often than 17 acknowledged communicating with 15. There are a number of possible reasons for this difference. Some people have better memories; others are more careful in responding to questionnaires. There is a slight tendency for the lower-status member of a communication pair to be more likely than the higher-status member to remember a transaction. Another interesting situation is the case of number 28. Several arrows come into 28's node but none go out. This individual was apparently a very high communicator. He did not have a very high regard for questionnaire surveys, however, refusing to indicate those with whom he communicated. In most cases, the arrows go in both directions. The degree to which such reciprocation exists provides a measure of the quality and reliability of the data.

It is immediately obvious from figure 6.5 that individuals vary greatly in the number of their regular communication partners. At one extreme are those who somehow keep in touch with nearly everyone else in the organization (the "stars"); at the other are the communication isolates with whom no one reported weekly

Figure 6.5 A Typical Communication Network in a Small Research
Laboratory

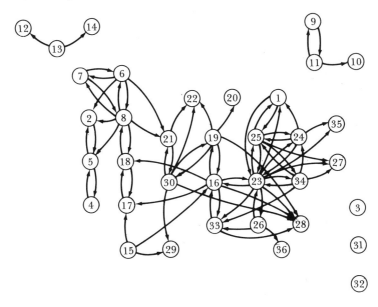

contact and who themselves did not deny this. Reflecting now upon the fact that personal contact within the organization had been the most effective source of information for the projects studied, it seemed apparent that such personal contact was more likely to have been with one of the communication "stars" in the network (for example, 19, 23, or 33 in figure 6.5). Thus it was decided to compare the "stars" with everyone else in terms of their degree of communication outside the organization through both documentation and long-term, well-established personal contact. In doing this, it was hypothesized that the communication "stars" would exhibit a degree of contact outside of their organizations that was significantly greater than that exhibited by their colleagues.

This hypothesis was first tested in a small chemical firm in the greater Boston area (Laboratory A). The hypothesis found support not only there but in several additional laboratories that were subsequently studied (figures 6.6 and 6.7). There thus existed in all of these organizations a small number of key people to whom others frequently turned for information. These key people ("technological gatekeepers") differed from their colleagues in the degree to which they exposed themselves to sources of technical information outside their organization. They read more, particularly the more sophisticated technical journals (figure 6.6). They do not appear to have read trade magazines and controlled-circulation journals any more than did their colleagues but they read the refereed journals significantly more. It will be recalled from chapter 4 that the readership of refereed journals was particularly low among the engineers on the projects studied. At that point, though, we were merely looking at the mean of a distribution, and the fact that the average engineer read very little did not imply that nobody read. In fact, a few people read quite a lot, and fortunately for their organizations they tend to be the same people to whom others come for information. The flow of information from the professional journals to the average technologist is a bit indirect, but it does occur. The gatekeeper

Figure 6.6 The Reading Practices of Communication "Stars." Data on Laboratory "M" from Frost & Whitley (1971)

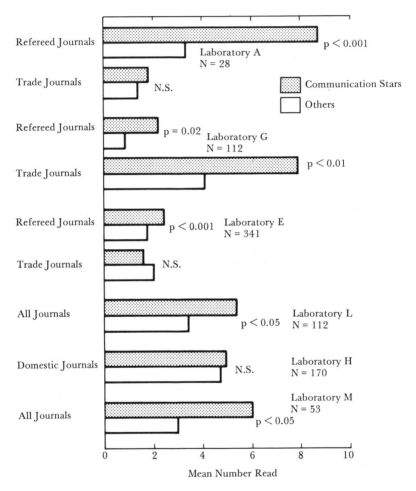

Figure 6.7 The External Communication Practices of Internal Communication "Stars"

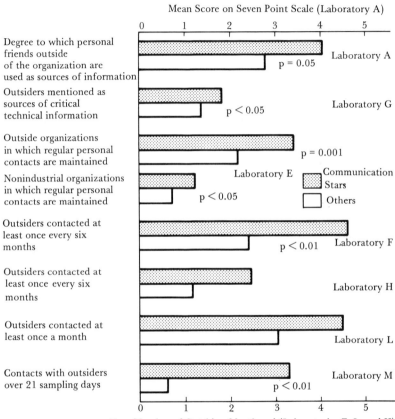

Mean Score on Seven Point Scale (Laboratory A)

Degree to which personal friends outside of the organization are used as sources of information — Laboratory A — p = 0.05

Outsiders mentioned as sources of critical technical information — Laboratory G — p < 0.05

Outside organizations in which regular personal contacts are maintained — p = 0.001

Nonindustrial organizations in which regular personal contacts are maintained — Laboratory E — p < 0.05

Communication Stars / Others

Outsiders contacted at least once every six months — p < 0.01 — Laboratory F

Outsiders contacted at least once every six months — Laboratory H

Outsiders contacted at least once a month — Laboratory L

Contacts with outsiders over 21 sampling days — Laboratory M — p < 0.01

Mean Number of Outsiders Mentioned (Laboratories F, L, and H)
Mean Number of Organizations (Laboratory E)
Median Number of Contacts (Laboratory M)

can understand at least a portion of the material published in the refereed journals and can then translate this information into terms that the average technologist can use. In a similar fashion, there are also a small number of individuals who have a broad range of personal contacts both within and outside the organization. It is important that the outside contacts are maintained on a continuing informal basis. The difficulty that was found earlier with external communications can probably be attributed to the fact that the contacts were very brief in nature and in most cases were prompted only by the development of a problem. As a result there was some difficulty in communication that should not occur in the case of people who come to know one another over time, with each developing an understanding of the other's work. This is the case with the gatekeeper who keeps his organizational colleagues in touch with current developments by means of his informal connections with the outside. The pattern of communication between the average technologist and the world outside his organization is shown in figure 6.8. Contact occurs most effectively through the gatekeepers in a two-step or multistep process.

Origins of the Concept

This is certainly not the first time that an indirect flow has been discovered between some medium and the eventual user of the information that medium supplies. A similar two-step process was discovered twenty-five years ago by Lazarsfeld, Berelson, and Gaudet (1958) in a study of voter decisions during the 1940 presidential election campaign. They found that information from radio and newspapers did not influence the average voter directly. Rather, it influenced a key subset of voters, or "opinion leaders" as they were called, who subsequently influenced the vote of friends and associates. It was subsequently found (Katz and Lazarsfeld 1955) that opinion leaders function in many of life's activities—from fashions to public affairs to cooking. Somewhat nearer our present concerns, opinion leaders have been found to

Figure 6.8 The Dilemma of Importing Information into the Organization.
Direct paths do not work (A) because literature is little used by the
average technologist and because the direct contact with outside persons
is ineffective. An indirect route through the technological gatekeeper
(B) has been shown to be more effective. Symbols next to incoming arrows
indicate the polarity of the correlation with performance.

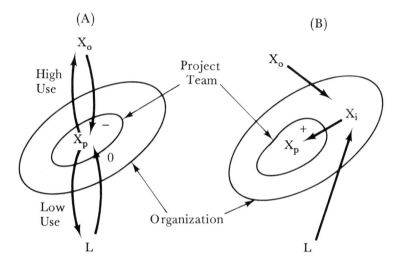

X_p = Project Team Member, in need of information
X_o = Person outside of the Organization
X_i = Organizational Colleague
 L = Literature

be influential in the propagation of such agricultural innovations as hybrid seed corn (Rogers and Shoemaker 1972) and in the introduction of a new drug to a community of physicians (Coleman et al., 1966). In both of these situations, the opinion leaders were found to be highly exposed to sources of information, either written or personal, outside their immediate community.

The phenomenon of the gatekeeper is not an isolated one. Rather it is one example of a much more general class of phenomena. There will always be some people who, for various reasons, tend to become more acquainted with information sources outside their immediate community. They either read more extensively then most or develop personal contacts with outsiders. A large proportion of these people in turn attract colleagues from within the community who turn to them for information and advice. This general pattern, which holds true in the R&D laboratory, has serious implications for management policy, which we will investigate further. First, however, we must deal with a theoretical issue that has caused some difficulty.

The Question of Causality

That it is the same set of people who are both more exposed to outside information sources and who attract organizational colleagues to them for consultation could lead one to suspect some sort of causal relationship between the two observations. The data, however, merely show that a correlation exists; they do not imply causality in any way. Furthermore, while the fact that a large number of colleagues come to an individual for information might conceivably drive him to outside sources in order to meet their requests, it seems unlikely that events generally follow such a sequence. That others turn to a person for information implies that there was some expectation that he had the needed information in the first place. Perhaps the opposite sequence holds true. Perhaps the individual who obtains information from the outside gains a reputation for "knowing the answers" and thereby attracts colleagues for consultation. This may be a bit more plausible than the first possibility, but it appears somewhat lacking in completeness. It posits far too simple a relationship. Certainly there are other variables that affect these relationships. A person's technical performance or his reputation for technical competence will certainly influence the propensity of others to turn to him for information. His administrative position within the organization might also either encourage or deter others in approaching him. The relationships among these variables have been explored in three studies. Frost and Whitley (1971), in their study of Laboratory M, performed a partial correlation analysis among measures of internal communication, external communication, and administrative position. They found strong correlations existing between the number of times an individual was approached by colleagues for technical information and both his journal readership and his external personal contacts. They also found strong correlations between the number of internal contacts and the administrative position of the individual (whether he was a section leader). When they controlled for administrative position, the correlation between internal contacts and journal readership decreased by 43 percent; the correlation with external personal contacts was unaffected. These results, although somewhat inconclusive, suggest the possibility that administrative status is the real determinant of the degree to which an individual is consulted by his colleagues and that external contacts are either caused by administrative status or are only spuriously related to internal consultation. This analysis is seriously lacking in two ways, however. First, it considers only the first level of supervision (section leaders), and while people may be more inclined to turn to first-level supervisors than working level engineers for information, they may be much less inclined to turn to higher levels of supervision. In other words, the relationship may be a curvilinear one. Second, another variable, technical performance, may be causally related to all of the other three variables. Frost and Whitley found that performance measured by "originality" is correlated with internal consultation and that this relation is unaffected by administrative status.

A recent study by Taylor (1972) produced remarkably contrasting results. When Taylor controlled for administrative status, the correlation between journal readership and organizational contacts actually increased by 45 percent. The relation between organizational and extraorganizational contacts was again relatively unaffected (10 percent decrease). Taylor had no measure of performance but agreed that it was probably the underlying cause of both internal consultation and administrative position. Technical competence both attracts others seeking information and inclines management toward promoting an individual to a supervisory position. The position of first-line supervisor makes a person more visible in the organization, so it too attracts those seeking information. It by itself is not sufficient to explain the gatekeeper phenomenon, however. In one study (Laboratory L) I repeated the same analysis with and without including supervisors and found individuals who met the gatekeeper specifications under both conditions. The actual relationship among the variables is most likely something like that depicted in figure 6.9. Many factors influence technical performance, but these very likely include exposure to information sources from outside the organization. Performance, in turn, helps an individual to develop a reputation for technical competence, which both attracts consultation from colleagues and leads to managerial recognition or reward through promotion. The supervisory position adds to the effect of the technically competent reputation and further attracts consultation. The gatekeeper is the individual who meets the requirements at both ends of the flow diagram and maintains a high level of communication both within and outside his organization.

Networks of Gatekeepers
As progressively larger organizations were studied, the resulting networks increased in complexity and analysis became more difficult. The network shown in figure 6.10 is but a single department

Figure 6.9 The Etiology of the Technological Gatekeeper Role

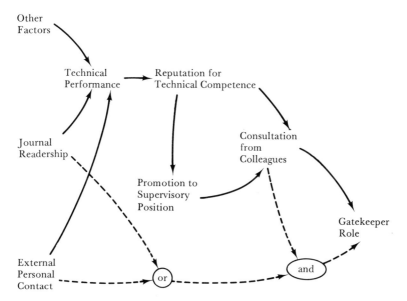

Figure 6.10 Typical Communication Network of a Functional Department
in a Large R&D Laboratory

In this study people were asked to indicate the names of colleagues from
whom they most often obtained technical or scientific information. Consequently, information generally flowed in a direction opposite to that
of the arrow, and the level of reciprocation is lower than will be found in
most of the network diagrams presented here.

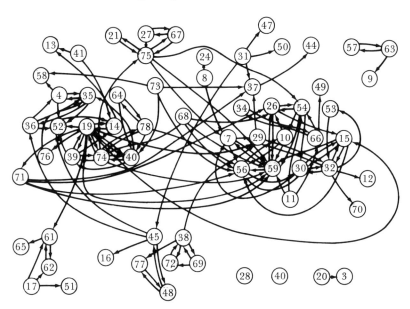

in an aerospace firm employing thousands of engineers and scientists. In my study of this firm, communication data were obtained from eight departments of about this same size. A network as complex as this is very difficult to understand; consequently I sought a technique that would both reduce the complexity and provide insights into the nature of the network and the reasons why a specific structure had developed. The technique I chose was a graph theoretic reduction into smaller, more cohesive components. Any communication network (or portions thereof) can be characterized according to the degree of interconnectedness or "connectivity" that can exist in a network (Flament, 1963). In the present analysis, only the degree of connectivity that Flament called "strong" will be considered. A strongly connected component, or a strong component in a network, is one in which all nodes are mutually reachable. In a communication network, a potential exists for the transmission of information between any two members of a strong component (Flament, 1963; Harary et al., 1965). For this reason, the communication network of Laboratory E was reduced into its strong components and their membership was examined.

When the departmental networks of the organization are reduced in this manner, two things become apparent. First, the formation of strong components is not aligned with formal organizational groupings, and second, while there were in each functional department from one to six nontrivial strong components, nearly all of the gatekeepers were found together as members of the same strong component (see, for an example, figure 6.11). In that organization (Laboratory E), 64 percent of all gatekeepers were distributed among eight strong components, one for each of the five technological and three scientific specialties. In each technical specialty, there was one strongly connected network in which most of the gatekeepers were members. The gatekeepers thereby maintained close communication among themselves, substantially increasing their effectiveness in coupling the organization to the outside world.

Figure 6.11 Departmental Communication Network after Reduction into Strong Components. Strong components are shown in brackets, and gate-keepers are shown by underlining with "G" superscript.

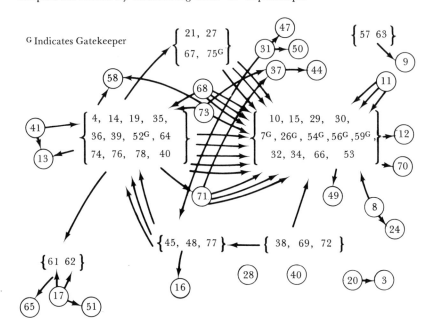

In figures 6.10 and 6.11 information flow is principally in the direction opposite to that of the arrows, and the reader will note the number of arrows entering the gatekeeper strong component. In other words, new technology generally enters the organization through one of the members of this component, is readily disseminated among them, and then spreads to the remainder of the organization through contacts that they have with colleagues outside the component. In this way members of the organization have a multiplicity of paths between them and sources of outside technology. This is presumably much more effective than a "star-satellite" system in which a number of individuals were associated with but a single gatekeeper and there was no communication among the gatekeepers.

Similar results were found in the other two large laboratories (H and L), except that in those organizations there appeared to be better coordination among disciplines (table 6.2). The majority of gatekeepers in these three laboratories are members of a single strong component.

Naturally there can be many variations on this theme. In some organizations a gatekeeper network will not occur at all; in others it will claim either more or fewer than the two-thirds average. The choice of a strong component to signify closeness of communication is purely arbitrary. Components of greater or lesser strength could have been chosen. However, it does make some sense, since among a mutually reachable set of nodes, some probability exists for a message entering one node to reach any one of the others eventually, particularly if it is actively sought by the eventual receiving node. Two additional considerations enter here. First, all of the analyses presented thus far have been based upon an average communication frequency of once or more per week. This fairly high level of communication further insures the effective functioning of a gatekeeper network. Second, there are additional levels of connectedness within the strong components, and gatekeepers are often even more closely connected to other gatekeepers with whom they share organizational affiliation,

Table 6.2 Distribution of Gatekeepers Among Strong Components in Two Organizations

Number of Nontrivial[a] Strong Components	Gatekeepers accounted for			
	Lab H		Lab L	
	Number	Percentage	Number	Percentage
0	0	0	0	0
1	23	88	8	62
2	24	92	10	77
3	25	96	11	85
4	26	100	12	92
5	26	100	13	100
6	26	100	13	100
7	26	100	—	—
8	26	100	—	—

[a]Membership greater than two.

physical location, or disciplinary background. An example of this can be found in Laboratory H where a single strong component connects twenty-three of twenty-six gatekeepers. The entire network for this laboratory is shown in figure 6.12. The organization performs R&D at six geographically separate locations, each of which is distinguished by a letter code in the figure. The seven larger centers (P, B, Y, S, K, R, and G) are each concerned with slightly different product lines; consequently, the centers generally draw from different disciplinary backgrounds although there is some overlap of disciplines among centers.

When this network is reduced, several strong components appear. The largest, however, accounts for nearly all of the gatekeepers (figure 6.13). Communication among the gatekeepers in this organization was extremely good despite geographic and disciplinary separation. Physical location and disciplinary interests show their effect, however, at a lower level in the analysis. The gatekeepers within centers have formed cliques that are more cohesive than the total strong component. In Center R, for example, there are seven gatekeepers. Six of these have formed a clique of order four. In other words, the maximum distance

Figure 6.12 Communication Network in a Geographically Dispersed
Organization

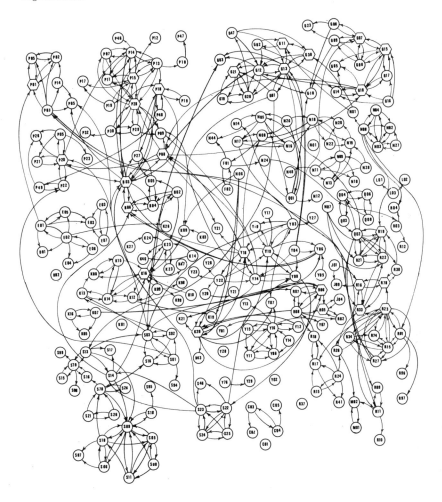

Figure 6.13 The Gatekeeper Network in a Geographically Dispersed
Organization

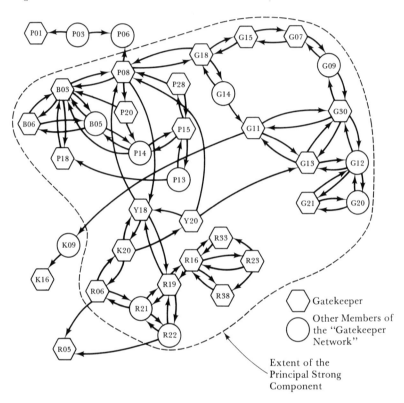

Gatekeeper

Other Members of
the "Gatekeeper
Network"

Extent of the
Principal Strong
Component

between any pair is four steps. Six of the seven gatekeepers of Center G have also formed a clique of order four. With the exception of PO3, the gatekeepers at Centers P and B have combined to form a clique of order four. While communication is good among all of the gatekeepers in this particular organization, it can be seen from the figure that it is especially good within each of the specific centers or when two centers have similar missions and draw from common disciplines, as in the case of P and B.

Given the constraints discussed earlier (reluctance of engineers to read, difficulty in direct communication with outsiders, and so on), it is difficult to imagine a system better than that of the gatekeeper network for connecting the organization to its technical environment. Even if one were to attempt to design an optimal system for bringing in new technical information and disseminating it within the organization, it is doubtful that a better one could be produced than that which, in many cases, already exists.

New information is brought into the organization through the gatekeeper. It can then be communicated quite readily to other gatekeepers through the gatekeeper network and disseminated outward from one or more points to other members of the organization (figure 6.14). Perhaps the most interesting aspect of this functioning of the organizational communication network is that it developed spontaneously with no managerial intervention. In fact, management of the different organizations were generally not aware that the network operated in this way.

Characteristics of Technological Gatekeepers
For numerous reasons, it might be advantageous to identify the gatekeepers in an organization. Because they are performing a vital role, management should be aware of the value of this activity and should see that the gatekeepers are appropriately rewarded. This does not imply formalization of the role, which seems unnecessary and could even prove undesirable. It merely means that recognition be accorded on a private, informal basis.

Figure 6.14 The Functioning of the Gatekeeper Network. New information
is brought into the organization by 1. It can be transmitted to 2, 3, and
4 via the gatekeeper network. It reaches its eventual users (indicated by
squares) through their contacts with gatekeepers.

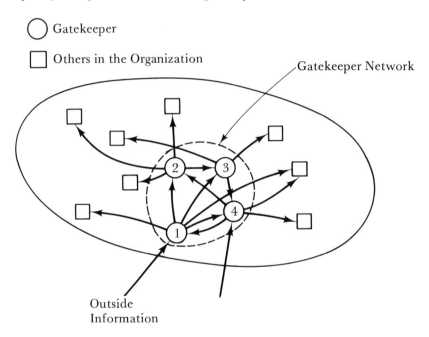

In addition, librarians and other purveyors of information might well be able to perform their functions better if they were to know the identity of the gatekeepers in their organization.

Given these needs, what can be done to identify the gatekeeper? Does one need to perform the sort of elaborate study that is discussed in this book? Fortunately, the answer to this question is no. The gatekeeper can be very easily recognized on the basis of three characteristics:

1. The gatekeeper is a high technical performer.
2. A high proportion (about 50 percent) of gatekeepers are first-line supervisors.
3. With a little thought technical management can generally guess accurately who gatekeepers are.

Technical Performance In addition to his indirect contributions, the gatekeeper is usually an important direct contributor to the organization's technical goals. Individual performance was measured in a variety of different ways in the different organizations that were studied. Almost any measure of individual performance in research and development can, of course, be challenged. Measuring individual performance is a very difficult matter. Recognizing this, it was decided to use a variety of different measures and check to see whether the results converged. Data are included from the three largest laboratories (E, H, and L). With the exception of patents and internal reports, all other measures clearly indicate the gatekeepers to be high performers (figure 6.15). They produce more papers for presentation at technical conferences, they are more frequently cited by management as "key people," and they receive higher ratings by peers and superiors.

Perhaps the most important performance measurement among the seven is the "key people" measure used in Laboratory E. Each chief engineer was asked to prepare a list of people whose loss would most seriously harm the organization in a technical sense. They produced quite an extensive list, naming about one out of every seven professionals in the organization. The proportion of

Figure 6.15 Performance of Gatekeepers

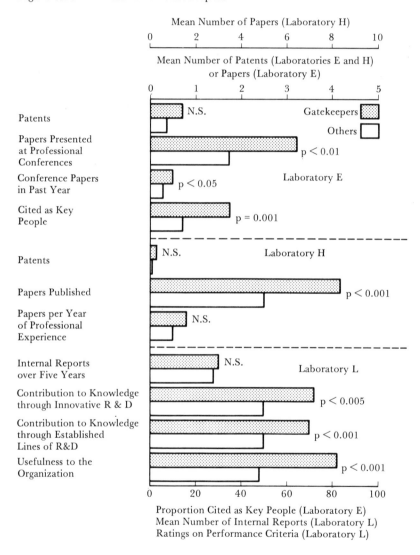

Mean Number of Papers (Laboratory H)

Mean Number of Patents (Laboratories E and H)
or Papers (Laboratory E)

Patents — N.S. — Gatekeepers / Others

Papers Presented at Professional Conferences — p < 0.01

Conference Papers in Past Year — p < 0.05 — Laboratory E

Cited as Key People — p = 0.001

Patents — N.S. — Laboratory H

Papers Published — p < 0.001

Papers per Year of Professional Experience — N.S.

Internal Reports over Five Years — N.S. — Laboratory L

Contribution to Knowledge through Innovative R & D — p < 0.005

Contribution to Knowledge through Established Lines of R&D — p < 0.001

Usefulness to the Organization — p < 0.001

Proportion Cited as Key People (Laboratory E)
Mean Number of Internal Reports (Laboratory L)
Ratings on Performance Criteria (Laboratory L)

gatekeepers on the list, in comparison, was almost 40 percent. Not all of the gatekeepers were considered key people in this sense. They were far more likely, though, to be so considered than the average member of the organization. The number of papers presented at scientific and professional engineering conferences provides a measure of the degree of recognition accorded to the gatekeeper from outside his organization. The gatekeeper also produced a significantly larger number of papers published in refereed journals, but in the study of Laboratory E, this number was measured only over the time period that the individual was in his current work group. The number of conference papers will be a function of an individual's age or career length too, but the gatekeepers were also found to have produced a significantly ($p < 0.05$) greater number of conference papers during the year prior to the study.

In Laboratory H, as in Laboratory E, the number of patents does not distinguish the gatekeeper. The number of papers published in refereed journals does. However, in this laboratory, the gatekeepers tended to be slightly older, with an average of four years more professional experience. For this reason the number of papers produced was normalized on the basis of professional experience. The gatekeepers still produced a greater number of papers per year of experience, but the difference is no longer significant statistically.

In Laboratory L, peer and supervisor ratings of contribution to the individual's research field and to his organization were adjusted to remove the effects of longevity in the laboratory and level of education. The technique used in adjusting scores was developed by Pelz and Andrews (1966) and is reported in appendix C to their book. From this set of measurements and from those performed in Laboratory E, a picture emerges of the gatekeeper as an individual whose technical competence is highly regarded both within and outside his organization. These seem to be essential ingredients in the formation of such a role. Before people will come to someone for information, they must feel that he has something to offer. They must respect his technical competence. Similarly, in order to obtain information, particularly outside one's organization, over the long run a person's competence must be respected. It is difficult to obtain information for very long from the same source without giving some in return, or at least presenting evidence of one's potential to provide worthwhile technical information. Competence is perhaps the most important quality for a gatekeeper. Consequently, it is impossible to establish gatekeepers without having them intimately involved in the work of the laboratory. This is what distinguishes the gatekeeper from the information specialist or information officer sometimes assigned to an R&D group. The information officer and gatekeeper are not identical functions, and they are not competing ones. They are really complementary functions. This is an important point and is one over which there seems to have been some misunderstanding. People have often interpreted the gatekeeper concept to mean that a person becomes intimately knowledgeable with the literature in a specific field and then directs his colleagues to appropriate sources. That, in a sense, is one function of an information officer. The gatekeeper may not be as knowledgeable about the various sources of information, but he does know far more of their content than the information officer does. In other words, while an information officer provides range, the gatekeeper provides depth. Moreover, the gatekeeper normally communicates content rather than direction to a source. He may do both, but his principal contribution comes by way of the translation that he can perform. He converts documentary information or information gained through personal contacts into terms that are both relevant to and understandable by the members of his organization. Because of this manner of functioning, it is almost absolutely necessary that the gatekeeper be an outstanding technical performer.

Tenure with the Organization Gatekeepers generally do not differ from their colleagues in terms of age (figure 6.16). In only one

Figure 6.16 Gatekeepers and Others Compared in Terms of Age

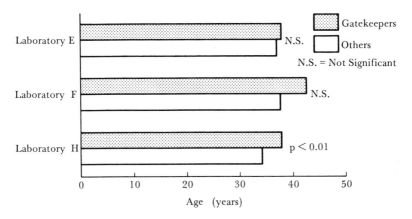

organization were they found to be significantly different from their colleagues and in that case, they were older. This is not surprising because there is no reason to believe that this role befits any particular age. Nevertheless one would expect it to take some years before an individual progresses into the role of gate-keeper for any particular organization. When gatekeepers were compared with their colleagues, however, in only one organiza-tion had they been around a significantly longer time (figure 6.17). These results, however, mask a threshold effect. It does take some minimum period before one can function as a gate-keeper. Seldom are gatekeepers found with fewer than five years in an organization, and never with fewer than two. It does take time to develop one's position in the organizational communica-tion network. This point is certainly understandable, but five years seems like an excessively long period. If an organization is fortunate enough to have hired an individual who has the po-tential of becoming a gatekeeper, it shouldn't have to wait five years to benefit from his services. There is probably little that can be done about the gatekeeper's external contacts and his propensity for reading journals. In those areas, he either has it or he doesn't. The organization can, however, do a lot about his internal communications. Here is where the techniques for in-tegrating new-hires, discussed in the last chapter, assume real importance. It is important to insure integration for everyone, but in the case of the potential gatekeeper, this responsibility should assume even greater prominence.

Professional and Organizational Status In all but one organization (figure 6.18) gatekeepers are more likely to hold a doctorate or medical degree. The one exception to this pattern is a small lab-oratory with only two people holding doctorates. The fact that this laboratory has the smallest proportion of doctorates tells something of its orientation in comparison with the other labora-tories studied. Its work was somewhat more developmental in nature and the possession of an advanced degree was of

Figure 6.17 Tenure with the Organization

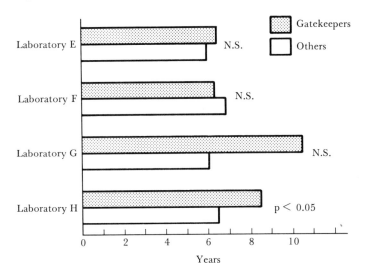

Figure 6.18 Proportion of Gatekeepers with Doctoral Degree

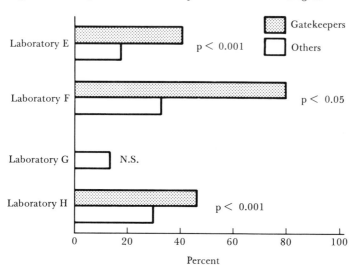

correspondingly less importance. In organizations in which doctoral degrees are important, gatekeepers will tend to have them and where they are not important, the gatekeepers will be correspondingly less likely to possess them.

Within the organizational hierarchy, gatekeepers are most frequently found at the first level of technical supervision (figure 6.19)—that is, they are usually the first level above the bench engineer or scientist and have roughly six subordinates. The first-line supervisor is perhaps in an ideal position to mediate between his small group (usually six or fewer) and the world outside the organization. Here again the problem of causality arises. These people might perform the gatekeeping function because they are first-line supervisors, or they might be first-line supervisors because the organization has rewarded their performance as gatekeepers. The reader is referred back to figure 6.9 for an interpretation of the relationships among gatekeeping, performance, and organizational position. We now know that gatekeepers are among the organization's highest technical performers, so they are probably rewarded by promotion for this performance, and for that reason a high proportion of them are first-line supervisors.

From the point of view of connecting small technical groups to the outside world, it appears that management is promoting the right people. The gatekeeper undoubtedly finds it easier than the nongatekeeper to assume leadership of his group. He is respected for his competence,[3] sought out by his colleagues for information, and has greater contact outside his organization. In addition, once he is in the position of supervisor, he becomes even more visible within the organization and is sought by people outside his group for information. This is evidenced by the fact that even when within-group contacts are eliminated, the gatekeepers have a significantly greater number of contacts from people outside their work group.

If gatekeepers function so effectively at first level of supervision, what happens when they are promoted again? Figure 6.20 provides at least part of the answer. Some continue to

Figure 6.19 Proportion of Gatekeepers at First Level of Supervision

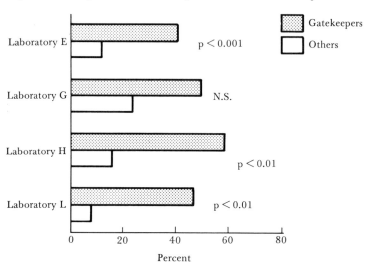

Figure 6.20 Proportion of Gatekeepers at Second Level of Supervision

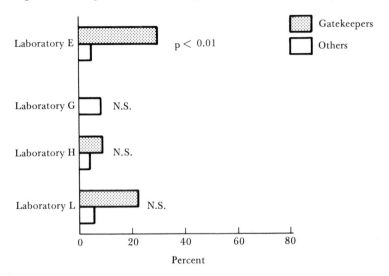

function as gatekeepers after promotion above first-line supervision, but they no longer constitute such a preponderance of the
population at second level as they did at first. Furthermore, as
they move up the hierarchy, they virtually disappear. Occasionally
one encounters a gatekeeper in the position of chief engineer or
chief scientist, or perhaps even research director. These occasions
are rare, however, and always involve a very rare individual whose
technical advice is still sought by those at the very bottom of his
organization.

First-level supervision, then, is the critical position. The gatekeeper is able to maintain effective contact with the working level
and provide needed technical information to people at those
levels. Do we then attempt to keep gatekeepers at this level and
refuse them promotions? Such a policy would be obviously self-
defeating. Just like everyone else, the gatekeeper wants to get
ahead in his organization. Data from two organizations show that
he does not differ from the average in this regard (figure 6.21).
Neither does he overstate the importance of promotion. The
gatekeeper, a high performer with well-established contacts outside his organization, is a potentially mobile individual. Frustrating
such a person can lead to but one thing—losing him. If a gatekeeper seeks promotion, and merits it, then he should be promoted. The organization, however, must have someone prepared
to assume his role by constantly developing new gatekeepers
through the techniques and policies that I have discussed.

Other Characteristics of Gatekeepers We now know that the gatekeeper is most often a technically competent first-line supervisor,
who has been with the organization an average of about six to
eight years. What else can be said to help in identifying or creating
gatekeepers? Perhaps the most obvious is that they are very busy
people, both within and outside their organizations (figure 6.22).
They participate more frequently in professional meetings and
conferences (a wise strategy on the part of their managements),
and they are more likely to be selected for special task forces for

Figure 6.21 Orientation of Gatekeepers Toward Advancement (Laboratory L)

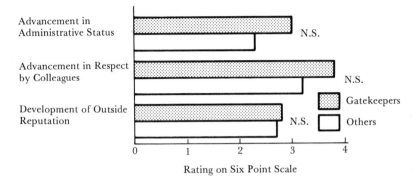

Rating on Six Point Scale

Figure 6.22 Activities of Gatekeepers

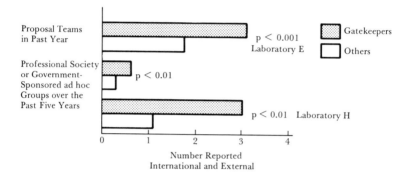

Figure 6.23 Work Orientation (Laboratory L)

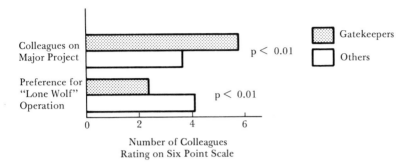

Number of Colleagues
Rating on Six Point Scale

Figure 6.24 Allocation of Time (Laboratory L)

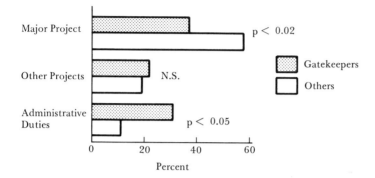

Figure 6.25 Creativity and Work Involvement (Laboratory L)

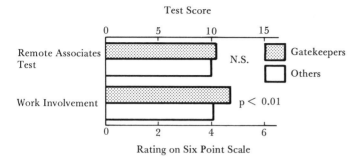

both their own organizations and for professional societies or
government bodies.

They tend also to prefer working with a number of people and
sharing responsibility for their projects (figure 6.23) rather than
operating as a "lone wolf." This makes some sense since they
would probably have to be somewhat people-oriented in order
to have very many colleagues approach them for assistance. They
also like to distribute their time somewhat over a number of activities rather than invest it primarily in a single major project (figure
6.24). They even devote a significantly larger portion of their
time to administrative activities. Finally, the gatekeeper is very
involved in his work, but on at least one test of creativity, the
Remote Associates Test (Mednick 1962), he does not perform
any better than do his colleagues (figure 6.25).

Recognition of Gatekeepers

The foregoing characteristics begin to spell out a very consistent
pattern of high performance, some orientation toward people,
a normal amount of ambition, and recognition of competence
from both within and outside the organization. We have saved
what is possibly the most important characteristic for last. The
gatekeeper is easily recognized. Once they pause to think about
the concept, managers generally know who their gatekeepers are.
Their guesses, from our experience, are generally correct. When
this was tested in one organization, the overlap between management's guesses and our data was in the region of 90 percent. In
other words, an elaborate communication survey is not necessary
to locate gatekeepers. All a good manager has to do is be sensitive
to the concept. He will find his gatekeepers. The only thing remaining is to reward them, organize information dissemination
around them, and if they are to be lost through promotion, re-
place them.

NOTES
1. The difference is significant statistically ($p = 0.02$).

2. The following sections draw heavily on the model presented by Katz and Kahn (1965).

3. The most important attribute a technical supervisor can have. See Farris and Andrews (1967).

7 STRUCTURING ORGANIZATIONAL COMMUNICATION NETWORKS I: The Influence of Formal and Informal Organization

Chapter 5 clearly showed the value of internal communication. Chapter 6 showed how the gatekeeper functions to bring technology into the organization. Now let us turn to the practical aspects of this situation. What can be done to improve the flow of internal communication, and what can be done to increase the contact of the laboratory staff with gatekeepers? The present chapter will look at ways in which communication with the organization can be structured, restructured, or improved, but first we must acquaint ourselves with some of the problems and barriers that impede internal communication.

BARRIERS TO PROMOTING INTERNAL CONSULTATION

Given the obvious benefits of internal consulting, it is curious and puzzling that it is so infrequently used. To probe deeper into this phenomenon, nineteen engineers in a large electronics firm were chosen for study. There engineers completed solution development records and reported all communication contacts for a period of three months. In addition they were asked to rank each of nine information channels (table 7.1) according to the following two criteria:

1. Accessibility. The degree to which one can attain meaningful contact with the channel—in other words just how easy it is to approach, obtain, or contact the channel (without giving consideration to the reliability or quality of the information expected).

2. Technical quality. Technical quality or reliability of the information obtainable from each of the nine channels listed (without giving consideration to the accessibility of the channel).

The second of these is a subjective evaluation by the engineer of the expected performance of an information channel. This should not (and, in fact, does not) agree with any of the more objective performance ratings reported elsewhere in this and other chapters. This is intended since it is the individual's *perception* of the situation that should determine his behavior.

Table 7.1   Information Channels Ranked by Nineteen Engineers

Literature	Books, professional, technical, and trade journals, and other publicly accessible written material.
Vendors	Representatives of or documentation generated by suppliers or potential suppliers of design components.
Customer	Representatives of or documentation generated by the government agency for which the project is performed.
External sources	Sources outside the laboratory or organization that do not fall into any of the above three categories. These include paid and unpaid consultants and representatives of government agencies other than the customer agency.
Technical staff	Engineers and scientists in the laboratory who are not assigned directly to the project under consideration.
Company research	Any other project performed previously or simultaneously in the laboratory or organization regardless of its source of funding. This includes any unpublished documentation not publicly available and summarizing past research and development activities.
Group discussion	Ideas that are formulated as the result of discussion among the *immediate* project group.
Experimentation	Ideas that are the result of test, experiment, or mathematical simulation with no immediate input of information from any other source.
Other division	Information obtained from another division of the company.

Similarly, accessibility is intended as a measure of the perceived cost associated with using a channel. Cost in this sense is defined very broadly to include physical effort, tedium, any loss of self-esteem, or other difficulty encountered in asking someone else for help or information.

The resulting rank orders of information channels for each of the two criteria provide subjective estimates of "channel cost" and "channel payoff" (or value). These rankings can be tested for correlation with rank orders based on the frequency of use determined from solution development record data. This technique makes possible the identification of the relative importance of various criteria in information channel selection.

THE DECISION TO USE A CHANNEL

Looking first at the influence of cost on channel selection, we find (table 7.2) a strong relation between channel accessibility and frequency of use. The first-order correlation (0.6) with the cost factor is more than double that of the quality factor (0.28). If there remain any doubts concerning the relative importance of quality or cost in determining this decision, a mere glance at the partial correlations should put them to rest. Holding perceived technical quality constant has almost no effect on the relation between accessibility and frequency of use. Holding accessibility constant drives the correlation between perceived quality and use virtually to zero. This, of course, means that perceptions of quality and accessibility are themselves slightly correlated, and the first-order relation between quality and frequency of use is illusory and appears only as a result of the mutual relation with the third variable, accessibility. In the minds of the subjects, there is apparently some relation between their perceptions of technical quality and channel accessibility, but it is the accessibility component that almost exclusively determines frequency of use.

FIRST SOURCE SELECTION

From the 154 information searches reported by the nineteen engineers, we are able to perform a second test of the channel selection hypothesis by looking at the channels that were approached first on each search (table 7.3). Once again, channel accessibility appears as the dominant criterion upon which selection is based.

Table 7.2   Criteria for Selecting an Information Channel

Correlation between Frequency of Use and:	Kendall Tau	Kendall Tau (Partialed)	Variable Held Constant
Perceived accessibility	0.67[a]	0.64	Technical quality
Perceived technical quality	0.28	0.03	Accessibility

[a] $p < 0.01$.

Table 7.3 Criteria for Selecting a First Information Source

Correlation between Frequency of Selection as a First Source and:	Kendall Tau	Kendall Tau (Partialed)	Variable Held Constant
Perceived accessibility	0.67[a]	0.61	Technical quality
Perceived technical quality	0.39	0.19	Accessibility

[a] $p < 0.01$.

Engineers turned first to the channel that was most accessible; perceived technical quality influenced this decision to only a very minor extent.

The engineers' behavior is, in this way, reminiscent of Zipf's (1949) "Law of Least Effort." According to Zipf's law, when individuals must choose among several paths to a goal, they will base their decision upon the single criterion of *least average rate of probable work*. In other words, to minimize his average rate of work expenditure over time, "an individual estimates the probable eventualities, and then select(s) a path of least average rate of work through these."

In the selection of information channels, the engineers in the study certainly appear to be governed by a principle closely related to Zipf's law. They attempted to minimize cost in terms somewhat broader than Zipf's, but if the psychological cost involved in asking someone for help with a problem is included as a form of effort, then engineers do attempt to minimize effort by turning first and most frequently to more accessible sources of information.

In doing this, however, they must either ignore probable work (searching through other channels upon failure to obtain the needed information through the first channel) or they are unable to estimate probable future work. If, in fact, engineers were to consider future effort in making their decision, there would be a negative correlation between the mean number of channels that must be used to gain the desired information after selecting one channel as the first source. In other words, for each time that a given channel is chosen as a first source, the number (including zero) of channels that had to be used before the information was in hand can be counted. The mean number of additional channels used can be computed for each type of channel. This mean score will serve as an index of the long-term effort (over and above that effort connected with gaining access) associated with each information channel. This computation was made using data from the nineteen engineers, and the near zero correlation that resulted indicates that engineers did not take this long-term effort into account in selecting channels. So we are left with engineers behaving according to a simplified version of Zipf's law in which they took only their immediately predictable effort into account and minimized that parameter in making their decision.

The implications of this finding are very important. Improving the quality or performance of a particular information service will not lead to increased use of the service. More investment in library holdings, for example, will be wasted unless this material is made more accessible to the user. Engineers will not be attracted to the library simply by improvements in the quality or quantity of the material contained there. The library must, in a sense, come to them. Microfiche may solve the library's storage problem, and it may make a greater amount and variety of information available to the user, but until it can be made as easy or easier to use than the traditional journal form, it will deter potential users.

Those concerned with management R&D—laboratory directors, librarians or administrators, or officials of professional societies— can control both dimensions of this problem to some degree. To some extent they can control the quality of the information that they make available. The laboratory director, for example, can recruit a competent technical staff to improve the quality of internal consulting. This by itself will not be enough, however. In addition, he must encourage their use as consultants by rewarding them for this activity and by encouraging engineers to go to them for information. He must make it known that no one will downgrade a person for seeking help and thereby take steps to reduce the psychological risk and cost in internal consulting.

THE EFFECT OF EXPERIENCE

One would expect that the degree of experience that an engineer has had with a given information channel would influence his perception of both the costs and the value associated with that channel. He might never use the library, for example, because of a misconception that it contains nothing of value or that it is really too difficult to find what he is looking for. Given some moderate amount of experience with library use, though, he might come to adjust his perceptions to be more in accord with reality. Similarly, in the case of interpersonal channels, increased experience should adjust perceived costs. The adjustment can mean either an increase or decrease in perceived cost and may be entirely random, increasing for some individuals or channels and decreasing for others. Should the latter situation hold, correlation analysis will fail to detect any systematic trend, and a zero correlation will be found between experience and perceived cost. If, on the other hand, experience brings about a systematic shift (most subjects shifting in the same direction with respect to most channels), correlation analysis can detect this phenomenon and will show a positive or negative correlation between experience and cost.

The data in table 7.4 demonstrate quite clearly that a strong positive relationship holds between the degree of experience an engineer has had with a given channel and perceived accessibility.

Table 7.4 Amount of Experience and Perceived Cost of Using Information Channels

Correlation between Degree of Experience and:	Kendall Tau	Kendall Tau (With One Variable Constant)	Variable Held Constant
Accessibility	0.72[a]	0.53	Frequency of use
Technical quality	0.56[b]	0.43	Accessibility
		0.39	Frequency of use

[a] $p < 0.01$.   [b] $p < 0.05$.

Of course, the engineer may simply refer more frequently to those channels he considers more accessible or easier to use. In other words, the direction of causality may be from cost to experience. He uses lower-cost channels more and thereby accumulates greater experience with them. To test this possibility and determine causal direction, the frequency with which the engineers used each of the nine channels was controlled and the relation between experience and perceived cost again tested (table 7.4). This reduces the original relationship somewhat, but reasonably strong correlation remains, thus indicating that the degree of experience that an engineer acquires with an information channel does tend to lower his perception of the cost of using that channel.

These results show that something can be done to improve information channel accessibility. As engineers gain more experience with an information channel, they modify their perception of it and see it to be more accessible. Therefore any steps to increase use through improving accessibility will be self-reinforcing. Laboratory managers who make their internal consultants more accessible will find that engineers will use the consultants more often, which will further enhance their accessibility. The same holds true for the library or any other form of information service or source.

THE ORDER OF SEARCH IN INFORMATION SEEKING

In addition to the frequency with which the nine information channels were used, data were obtained from the nineteen engineers on their search order during 111 information searches. Engineers indicated on forms the sequence of channels that they consulted each time they needed information. If we look now at the frequency with which channels appear as first and last sources in a search sequence, some insight may be gained into the cause of engineers' apparent reluctance toward internal consulting and the exact nature of the costs associated with this particular source of information.

This comparison of the order in which information channels were contacted by an engineer with a specific information need reveals some rather interesting differences. Figure 7.1 compares the distributions of first and last choice (channel approached first or last) in 111 information searches with the distribution of choices that would be expected from observation of the engineer's overall information-search behavior.

If an engineer's probability of selecting particular channels as his first source in an information search were governed by the same factors that determined the overall frequency with which he used each of the channels, there would be no significant difference in the proportions shown in the first[1] and second bars of figure 7.1. This follows directly from the fact that the expected frequencies are based on the observed pattern of the engineers' overall use of the four information channels. That significant differences exist is an indication that the selection of a first source was influenced by factors unique to the first-choice selection process itself. This proposition is further supported by comparing the first and third bars in each portion of the figure. Comparing the frequency with which particular channels appear as the final source in an information search with their expected frequency of use reveals no significant difference. Final sources are distributed over channels with about the same frequencies that would be expected from observation of overall information use patterns.

The engineers in the study approached the laboratory's technical staff far less frequently as a first source than their overall use would suggest. Engineers' reluctance to approach the technical staff of their own organization is manifested quite strongly in the sources to which they turned first for information. In other words, while other evidence indicates a somewhat general reluctance to approach the technical staff of their own organization for information, figure 7.1 shows this reluctance to be magnified in the case of the first source approached. It should be noted that the category "personal contact within the laboratory" includes anyone outside the project team. Had it not included members of the

Figure 7.1 Use of Information Channels as First and Last Source (111 Information Searches)

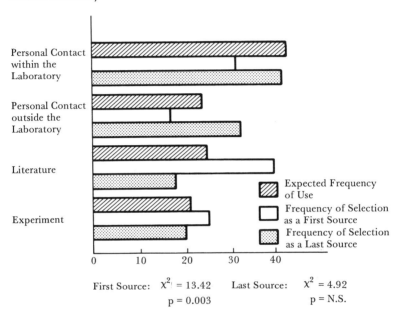

First Source: $\chi^2 = 13.42$ Last Source: $\chi^2 = 4.92$

p = 0.003   p = N.S.

searching engineer's own functional group, the numbers in both the expected and observed bars probably would have been even smaller.

A second interesting point manifested in the figure is the tendency to use the literature more frequently as a first source than would be anticipated. One might expect engineers to contact colleagues first in order to locate what they needed in the literature and use the literature at later points in the search process. In fact, they first consulted the literature and only then turned to colleagues. From interviews, we were able to learn that they are often consciously aware of this practice and follow it in order not to be caught in a position in which they appear too uninformed or naive when they approach a colleague for help.

THE NATURE OF THE CONSULTING PROCESS

There are many ways in which one might view the process of consulting within an organization. One possible way of looking at it is as an exchange transaction (Blau, 1963) in which both participants have costs and gains. The information seeker gains the information he needs to solve his problem, but in doing so, he has had to admit the superior competence of the consultant. This price can be particularly high in a bureaucratic organization where colleagues are all, to some degree, competing for advancement. The consultant, on the other hand, gains some prestige as the result of being asked for information, but he also incurs the cost of being distracted from his own work. Distraction from his work is particularly costly because the reward systems in most laboratories are structured to recognize only direct contributions to organizational goals. Little or no account is taken of such indirect contributions as consulting.

The costs in the consulting transaction are very often perceived to be far greater than the benefits. In the case of the consultant, cost rises with each successive consultation, while the value received in terms of ego rewards rapidly reaches diminishing returns. In his study of a government regulatory agency, Blau (1963) described this situation of diminishing returns aptly: "All agents like being consulted, but the value of any one of very many consultations became deflated for experts and the price they paid in frequent interruptions became inflated." Homans (1961) stated this principle in more precise terms as one of his propositions of human exchange: "The more often a man has in the recent past received a rewarding activity from another, the less valuable any further unit of that activity becomes to him." In far more homely terms, *an engineer who very frequently seeks consultation will soon wear out his welcome,* unless, of course, he has something substantive to offer in return. Information "sinks" are usually short-lived. Unless a person is able to give information in return, the tendency is to cut him off before very long. It will be seen later that very strong communication bonds develop among the best sources of information in the organization. They have something to exchange.

On the other side of the bargain, the information seeker also sees the transaction as a costly one. For him, perceived cost very often exceeds perceived value from the very outset. This is particularly true when, as in the case of strangers, there is even a very remote possibility in the information seeker's mind that the potential consultant will respond in an ego-threatening way. Moreover, even when initially perceived costs are not so high as to preclude consultation, they soon increase if repeated consultations are necessary. This results from the fact that a person's reputation in an organization can be severely damaged if he is too often seen in the role of an information seeker. Blau pointed out these same two aspects of the cost to the information seeker.

Asking a colleague for guidance was less threatening than asking the supervisor, but the repeated admission by [the information seeker] of his inability to solve his own problems also undermined his self-confidence and his standing in the group. The cost of advice became prohibitive if the consultant, after the questioner had subordinated himself by asking for help, was in the least discouraging—by postponing a discussion or by revealing his impatience during one.

To an engineer these costs are at least equal to, and are perhaps many times greater than they were for Blau's agents (compare Shepard, 1954). An engineer's prestige among his colleagues is founded to a great degree upon an almost mystical characteristic called "technical competence." To admit a lack of technical competence, especially in an area central to the engineer's technological specialty, is to pay a terrible price in terms of lost prestige. Furthermore, the problem is exacerbated when the consultant resides in the same organization as the individual who is seeking help. The individual must not only admit that he needs assistance in an area in which he is supposed to be technically competent, but he must continue to work with the consultant. He may actually damage his reputation in the organization, or at least be inclined to fear the possibility.

Strangely enough, this possibility and the resulting inhibitions do not exist in the case of personal contact outside the organization. The cost to the individual of consulting an outsider is often less than the cost of turning to an organizational colleague. The loss of prestige cannot be nearly as great when the relationship is brief and the chance is slight that any detailed knowledge of the transaction will ever reach back to one's own organization. The engineer can often evade the loss of prestige in this situation. Since the outsider has no knowledge of the person's organizational reputation or of the specific technical content of his responsibilities, the engineer can easily excuse his lack of knowledge by pretending to be an "expert in something else" who needs some help in "broadening into this new area." By such a stratagem, the engineer can disguise his lack of competence in an area in which he is supposed to be knowledgeable and run little risk that anyone in his own organization will ever learn of his game. The ruse would, of course, not work as well when consulting with someone within his own laboratory since the risk of being exposed is far too great.

The cost to the external consultant of being distracted from his own work is far less than it is for the internal consultant. To be sought after by someone from another organization is indeed flattering; it provides a considerable boost to one's prestige to have it known that people from other organizations come for technical advice. For the insider this fact is usually not evident. Someone approaching from within the organization is not as visible as a visitor entering from outside, and no one is aware of why the insider is talking to the consultant. For all the supervisor knows, they may be fishing companions talking over last week's (or next week's) catch. The reward is less and the cost greater. The person who consults with outsiders has an organizational blessing on his activity; if this were not the case, the organization would probably have discouraged the visitor in the first place. The internal consultant, on the other hand, pays the full price of spending time at the expense of his organizationally assigned responsibilities.

Interviews with the nineteen engineers who had reported on the cost of using different information channels are very revealing in this sense. They shed considerable light on the exact nature of the costs encountered in the consulting transaction. One of the engineers described the cost of approaching a previously unknown member of the organization: "I think people, being human, are somewhat reluctant to go to a person that they don't know. Either you are afraid you are going to look like a 'schnook' when it's all over or you are afraid that this guy may not have enough time. I think everybody goes through this ever since they were kids."

The reluctance to communicate openly is observed not only among newer employees but also among those who have been with the organization for several years. Their perception of the problem is somewhat different from that described by their younger colleagues. An engineer who had been with the organization for several years described how the cost of seeking information can increase with time in the organization: "When you come into a place as a junior engineer, you ask anybody anything and they accept the fact that you are real dumb. After you have stayed awhile, you ask fewer people fewer things. There are certain things that are expected of you and you are a little bit querulous [sic] about displaying ignorance. I think this is a very human reaction. You use some discretion in digging out information."

STRATEGIES FOR REDUCING COST
The interviews were especially valuable in turning up several cost-reducing strategies that engineers have developed. They also point the way toward actions that can be taken to remove some of the barriers to freer consultation in R&D organizations. Three strategies emerge from the data.

Oil on Troubled Waters
One engineer described a typical strategy as "putting oil on troubled waters." When he required technical information, he pursued the following course:

You have to start some place so I feel it's better to start off by asking stupid questions so you can build up to the intelligent ones, and when I feel it's a stupid question, I'll just simply say that. I don't know any better so I'm going to ask a question and when I get to the obvious, tell me so. If I thought I knew the answer, I wouldn't ask the question. And this perhaps is a way of putting oil on troubled waters. It tends to make the person to whom the question is addressed a little more receptive to the question. You don't want to use this approach too often; pretty soon people will tell you "Smarten up Charlie."

The engineer who uses this approach anticipates any potentially derogatory reaction by the individual to whom the question is addressed. By denigrating himself, the inquirer deprives his colleague of an opportunity to make a derogatory comment and simultaneously evokes empathy from the person being questioned. This approach lacks some of the additional benefits that can be derived through the next strategy, but it may sometimes be effective in reducing the threat faced by the information seeker. Of the three strategies, this is probably the poorest in overall effectiveness and is too much a function of individual personality to lend itself to effective encouragement by management.

Literary Preparation

The evidence in table 7.5 indicates that engineers often use the literature to prepare for face-to-face communication or to avoid oral communication completely in certain situations. As one engineer said, "If I were in a situation where I didn't know anyone, I would most likely ask fewer questions and would try to find out the answer on my own. This would be a case where literature would be most frequently used. I would like to know a little more before I asked a question." Another agreed: "Literature is frequently used to keep other people from knowing how little you know."

Very often the information seeker will find it difficult to formulate his question properly. To avoid appearing too ignorant, and thereby incurring a very great cost, he attempts to improve his background in the area by using a source that cannot directly challenge his ignorance. To the extent that he is successful, he will then be able to frame his questions in a manner that will not leave him too open to challenge. This strategy undoubtedly has the additional benefit of improving the quality of the consultation when it finally occurs.

Neutral Social Interaction

The third and most commonly cited technique used in interpersonal communication to reduce potential threats to an engineer's ego is to restrict communication to those engineers who are also known socially. The term *socially* as used in the present context refers to several types of interaction observed in the laboratory. Interactions may be recreational, such as playing bridge during lunch or discussing a weekend trip, or they may be an outgrowth of work activity, such as borrowing tools or test equipment. In those interactions evolving from work situations the topics discussed are neutral; they do not involve the exposition of technical knowledge, or lack of it.

These neutral social interactions, as well as the recreational activities, serve the important function of developing interpersonal understanding. Engineers indicated over and over again that understanding between colleagues was a prerequisite for effective technical communication. According to one:

I would say that establishing an initial understanding is the biggest block to communication. For instance, I am thinking of a fellow over at another division I got to know pretty well. He was a circuit designer. I'm in microwave. There is quite a gap between the two. He was a very good circuit designer and I get into these circuit problems every once in a while and someplace along the line I got to asking him questions, some of them were very stupid questions. He would answer these, spending a good deal of time and effort, so after I got confidence in talking to him that he would be sympathetic and answer as fully as he could, then I would go ask him anything.

and

I think this understanding has to come through personal contact, whether it's through bridge games during lunch hour or contact in the laboratory such as borrowing a person's equipment and returning it, or perhaps having something in common with the person like enjoying skiing or going to town meetings together or something like this.
    Sometimes you will ask someone a question and they will just beat around the bush. They don't know the answer either but they don't want to admit it because they are supposed to have a higher rating than you. If you know the person well enough, then you can joke about the whole thing. By well enough, I don't necessarily know them socially where you go out on the town together every weekend, I mean just perhaps socially like playing cards during the lunch hour. Bridge is a big out; it is a big social function as far as I can see.[2]

In a potentially tense situation when either a naive question is asked or the individual to whom the question is addressed does not know the answer, a loss of face is averted by joking about the problem. Goffman (1955) described this mechanism for correcting a loss of face:

Some classic ways of [re-establishing a situation] are available. On the one hand, an attempt can be made to show that what admittedly appeared to be a threatening expression is really a meaningless event, or an unintentional act, or a joke not meant to be taken seriously, or an unavoidable, "understandable" product of extenuating circumstances. On the other hand, the meaning of the event may be granted and effort concentrated on the creator of it. Information may be provided to show that the creator was under the influence of something and not himself, or that he was under the command of somebody else and not acting for himself. When a person claims that an act was meant in jest, he may go on and claim that the self that seemed to lie behind the act was also projected as a joke. When a person suddenly finds that he has demonstrably failed in capacities that the others assumed him to have and to claim for himself—such as the capacity to spell, to perform minor tasks, to talk without malapropisms, and so on—he may quickly add, in a serious or unserious way, that he claims these incapacities as part of his self. The meaning of the threatening incident thus stands, but it can now be incorporated smoothly into the flow of expressive events.

Both individuals are thus able to continue interacting in an otherwise potentially awkward interpersonal situation; on the other hand, when understanding is absent between the parties involved, blocks to future communication are far more likely to develop. This is an important point and cannot be overstressed. If the individuals know and understand each other, it is much easier to recover from an awkward situation. It is the implicit awareness of this mechanism that makes the engineer more willing to approach someone with whom he has become acquainted.

When two people are acquainted, they are also able to communicate more effectively. The understanding that develops between engineers through their social and work contacts is therefore important not only in encouraging communication but also in increasing its effectiveness. If one individual is familiar with another's background, he is better able to tailor his responses to the other's abilities. In reference to this, one engineer said: "Well, I think that in talking to somebody you know, you're familiar with what the person has done before and he probably can speak the language that you understand; whereas, if you talk to somebody whom you don't know, he might not explain it in a language you are used to hearing. Somebody you know understands what

your background is and you understand what his is and you usually can get together easier."

Considering the three general strategies to overcome interpersonal barriers to open communication, it would seem that the
third approach—developing greater understanding through increased social contact—will in the long run be the most effective,
as well as the most easily encouraged by management.

Other Cost Reduction Measures

The cost reductions discussed up to this point certainly help in
providing a better understanding of the internal consulting process
but produce little in the way of direct action that management
might take to improve the situation. What can be said about direct
management intervention? Facing this same problem of overcoming the reluctance of employees to consult with their organization colleagues, both Blau (1963) and Shepard (1954) suggest an
interesting possibility. Both of these analysts suggest an approach
in which technical discussion sessions function to maximize gain
to the participants while minimizing cost. Blau, for example, says:

The recognition of both participants in a consultation that one
provided an intellectual service to the other raised the status of the
consultant and subordinated or obligated the questioner to him.
These were the inducements for the consultant to give advice, and
simultaneously, the cost incurred by the questioner for receiving
it. Discussion of interesting problems, on the other hand, were not
recognized as providing a service to the speaker, and he did not
start them because he experienced a need for *advice*. Manifestly,
both he and the listeners, who sometimes commented, participated in these discussions because they were stimulating. The fact
that they facilitated his solving of problems was disguised from the
speaker as well as from his listeners; this was a latent function of
such discussions.

In the absence of awareness that a service was furnished, no need
existed for the speaker to reciprocate for the help he did, in fact,
obtain. He did not subordinate or obligate finding an audience,
since interest in the problem and its solution supplied sufficient
motivation for listening. This constituted the major advantage of
consultations in disguise over direct consultations. We find . . .

that the extraneous factors that motivate an interaction pattern
that is not intended to, but does, fulfill a given function make it
more efficient than a different pattern intended to, but does,
fulfill a given function make it more efficient than a different
pattern intended to fulfill this same function. Only a service in-
tentionally rendered creates obligations, which make it costly.

Shepard, in describing the activities of a project group in a
university-affiliated laboratory, has this to add:

The exchange of technical information was a social act, with a
significance in terms of interpersonal relations in addition to its
significance as part of a body of knowledge. The greatest respect
was reserved for those who had proved themselves competent in
solving the most sophisticated problems. The provision of technical aid to those who had problems to solve was at once a sign of
solidarity and a contribution to the system of reciprocated acts
which kept the currency in circulation. . . .
    A series of exchanges in which everyone contributed technical
information of equal value would result in the same relative
statuses at the end of the series as at the beginning, but each
member would be wealthier in terms of technical knowledge than
before, and his status related to members of other groups would
be increased.

Technical discussions in which all participants are able to make a
contribution would circumvent the dysfunctional aspects of social
exchange underlying the consulting process. Seminars that included engineers from several projects and functional areas could
be organized around topics so that all participants would appear as
equals. In this way, the status differential inherent in the consulting relationship can be avoided. While members of the more
esoteric functional staff units (from which consultants will be
more frequently drawn) will bring information of the more elegant
variety concerning physical theory, mathematical techniques, and
recent state-of-the-art advances, those engineers assigned to more
developmental projects bring an equally important contribution in
terms of specific new problems and applications. Since the inherent value ascribed by the technological community to these
two types of information is somewhat out of proportion, manage-

ment must assume responsibility for redressing the imbalance. The importance of information concerning the nature of current and future applied problems must be stressed, as must the potential contribution of project team members to the work of functional staffs. The importance of the feedback loop that supplies problems to the functional research groups cannot be overemphasized.

Periodic seminars organized around specific problem areas or specific technologies in which management feels that the laboratory has particular competence hold great promise for improving the flow of information between technical specialists and project members. This is not to say that the device will entirely eliminate the need for bilateral consultation. Actually it will probably increase the amount of bilateral consultation and improve its quality. It will do this in three ways. First, the seminars will increase the visibility of staff specialists. Very often, a project engineer will contact an outside source for information simply because he is unaware of the expertise available within the organization. It is astonishing to learn how generally ignorant the average member of an R&D laboratory is of work going on around him. Simply increasing the general awareness among members of what is going on in the laboratory is almost certain to enhance overall performance in the long run.

Second, the exchange process taking place within the seminar discussions diminishes the one-sided nature of later bilateral exchange. This results from the fact that the project engineer has demonstrated his ability to make a positive contribution to the knowledge of the specialist. When he later approaches the specialist for help, he is first of all not unknown and, more important, he is recognized for his particular brand of competence. He is seen as a potential contributor and not as a pure information sink.

The third (and probably most important) result will occur through the creation of strengthened social bonds among the participants in the seminars. We have shown quite clearly the importance of knowing beforehand the individual from whom a person seeks help or advice. If the seminar program achieves nothing else, it will acquaint members of the laboratory staff who might not otherwise meet. This simple device of developing acquaintances will reduce the deterrents to bilateral communication within the organization. It will provide both potential consultants and their clients (and the roles are certainly interchangeable) with invaluable information about the other's work interests, abilities, and needs. We have seen the importance of this sort of knowledge and the requirement for informal contact not necessarily related to the individual's task in laying the ground for effective consulting when it becomes needed.

In addition to all of these arguments, empirical support can be mustered for the seminar plan by referring once again to the study by Shilling and Bernard (1964) of sixty-four biological laboratories. Shilling and Bernard measured the extent to which informal technical discussion groups had formed in each laboratory. Specifically, they asked each respondent, "Are you a member of any group that informally discusses research?" They summarized the general finding in the following manner:

On the basis of a 50 percent participation criterion (50 percent or more of the respondents answering in the affirmative) it would seem that participation in group discussion is a custom in the scientific community; for example, in the median laboratory 60 percent of the scientists did participate in such groups. However, the great variability and the composite nature of the distribution suggests that the composition of the total set of laboratories may be blurring the picture. . . .

Participation in discussion groups is clearly not part of the culture of the government laboratories; nor is it of the industrial laboratories. In none of these two types of laboratories did as many as half of the scientists report discussion group participation. In the other three types (private university, public university, and private research institute), well over half did.

The median private university and the median private research laboratory tended to show a larger proportion of their staff participating in such groups than the median public university laboratory. But among all three, it could clearly be said that group discussion was an established custom.

Because of the more bureaucratic organization of the industrial and government laboratories, it may be that taking time off for group discussion has not yet become recognized as a valuable scientific activity. In the universities, such discussion is part of a long academic tradition.

Shilling and Bernard, it will be remembered, found statistically significant direct correlations between the extent to which laboratory personnel participated in these group discussions and seven of their eight measures of laboratory performance.

Industry has often attempted to emulate and simulate the university atmosphere in dealing with its scientists and engineers. For example, many corporations have built campus-type facilities. Kaplan (1965) has commented at length on this phenomenon and points out they have for the most part operated on a false set of assumptions that failed to recognize the differences between engineers and academic scientists. Perhaps we now have a situation in which industry could well profit by emulating an academic tradition.

In addition to the seminar program, management must take other steps to increase the general awareness of various laboratory activities and of information that is available within the laboratory organization. This might involve the publication for in-house use of directories of research projects and personnel or of brochures similar to those often put together for marketing purposes. Some experimentation is obviously required. One good possibility would involve the formation of purely technical review panels. Frequently an R&D project will be subjected to a management review, in which a panel comprising members of the laboratory's management is presented with evidence of the project's status in terms of costs, schedule, manpower, and so on. An analogous technical review system might be established in which the management panel is replaced by one composed of several of the laboratory's leading technical specialists in appropriate areas, and the focus of the review is upon progress on the project's key technical problems. In this way, the staff specialists are brought directly into contact with the project team.

Of course, in order to function properly, the project members must feel that they will not suffer for having revealed their problems. This is an essential point, and the success of the system rests entirely upon it. Should they feel that the revelation of problems will be held against them as evidence of incompetence or recalled at the time of their next merit review, they will naturally conceal problems and frustrate the system. Project members must feel that they can trust the panel. The creation of trust is going to be difficult to accomplish, but it is absolutely essential to the success of the panel system. There are at least two steps that management must take at the outset in order to create the proper atmosphere for the functioning of technical review panels. First, members of management above first-level supervision must be excluded from the panel and from panel review sessions. Second, project engineers themselves must have a strong hand in the selection of panel membership. The exclusion of management above the first level is an absolute necessity if project members are to feel that their future with the organization will not be affected by their candid revelation of technical problems. The second point supplements this. Election by project members would prevent the suspicion that the panel was a surrogate for management. On the other hand, management must have some voice in this selection process, if for no other reason than to insure against the election of friends who have no real claim to authority in the technical areas of concern and, of course, to exercise some control over the allocation of manpower in the laboratory.

Seminars are not a complete solution, however. If the laboratory is organized in such a way that the project groups seeking information and the staff or consulting groups supplying it are readily differentiated and recognizable, then a status distinction will almost necessarily develop and interfere with the goal of the seminar program. Should this occur, the confrontation between groups will develop the same asymmetrical relationship that was discovered for individuals. This is a very serious limitation and one that should be countered.

One solution to this problem lies in the form of organizational structure. "Project" and "staff" distinctions can be avoided to a large degree through proper structuring of the organization. Matrix organization accomplishes this quite effectively. Nearly everyone in the matrix organization can be associated with a staff group as well as with one or more projects. The distinction between the two becomes ambiguous, thereby reducing the likelihood that a status distinction will arise. Of course, matrix organization is not a complete solution either. Efforts must be directed toward changing the attitudes of the people involved in such a way that they will weigh the values of problem-related and solution-related information equally. The project members must be viewed not merely as information seekers but also as suppliers of valuable information on the nature of the problems to be solved. The technical staffs must be convinced of the value of this problem-related information in keeping their work relevant to practical needs. This is not a goal that can be accomplished overnight, but it is certainly a long-term one that every technical organization should strive to achieve. To whatever extent possible, the organizational climate must be structured in such a way that problem-related information and the communication of problems are highly valued.

An important drawback of the seminar approach lies in the general complaint among R&D personnel that far too much of their time is taken up by meetings already. In many organizations, this is probably a legitimate complaint. The addition of a seminar program in such circumstances would simply evoke the predictable response of poor attendance. For this reason, to be at all effective, any group interaction that is planned must be far more goal oriented than the usual academic seminar. The goal must forcibly draw attendants away from competing activities and provide a continuing rationale for the discussion sessions.

Ad-hoc groups are formed for many purposes. Short-term projects and task forces (see Galbraith, 1969, for a discussion of the task force as an integrating mechanism for groups) are but two obvious possibilities. The ad-hoc teams provide a context in which people can become acquainted while still maintaining contact with their more permanent groups. Even after dissolution of the ad-hoc group, the relationships formed in it will persist for a time. In this way, a greater proportion of the members of an organization will become acquainted, and the barriers toward subsequent bilateral consultations among them will thereby be greatly reduced. The development of acquaintances shows great promise for improving communication, particularly in large organizations.

Three additional measures can be taken to encourage the use of bilateral consultation. First, the organization's reward system should be restructured to reflect the importance of this activity. In other words, technical staff specialists should be rewarded for assisting the members of projects with their problems, as well as for work directly related to their own problems. Second—and this will be taken care of in part by the seminar program—engineers should be made aware that seeking consultation will in no way reflect on their own competence but will be recognized as an attempt to increase their own competence. Both of these measures attack the costs associated with the social exchange transaction. In addition, key individuals should be identified who are not only knowledgeable in particular areas but who are capable of translating technical information to match the project engineer's needs, and their identity should be publicized in the laboratory.

THE INFLUENCE OF ORGANIZATION ON COMMUNICATION PATTERNS

The subject of social relations and their influence on communication patterns brings up the more general subject of factors that influence the development and structuring of technical communication patterns in an organization. To determine the ways in which communication patterns develop and the factors that determine their structure, studies were conducted in a number of R&D laboratories of various types.

In these studies, the structure of an organization's communication network was determined by asking people to report all of their internal communication contacts. (The exact methods used are described in more detail in chapter 2 and the appendixes to this book.) The pattern of these contacts was then compared with, for example, the structure of the formal organization in the laboratory as shown on the organization charts. In addition the structure of what might be called the informal organization was obtained by asking people to name those with whom they most frequently ate lunch or with whom they interacted socially outside the work setting.

The Networks

Figure 7.2 shows a fairly typical communication network taken from a department in a large aerospace firm. Superimposed on the network are measures of the formal and informal organization. The department comprises five sections with a section head in charge of each. The chief engineer for the department is shown in the center circle. The section heads are shown in the next larger circle (one section had no formally appointed head at the time of the study), and section members are in the five truncated wedges. Communication contacts at a frequency of once or more per week are shown by arrows, and lunch or social contacts are shown by dotted lines. Communication is heaviest within sections, with very little communication among sections. In this department, the structure of the formal organization is the more important determinant of communication. The direction of causality is clear in the case of formal organization. It is very doubtful that communication by itself very often causes the creation of a formal organization. On the other hand, assigning people to the same organizational group (often through the intervening variable of physical location) very often results in technical communication occurring among them.

With informal organization a similar situation exists. When there is social contact between any two individuals, the probability of technical communication is significantly higher than when no social contact exists. In another department, shown in figure 7.3,

Figure 7.2 Communication Network in a Typical Department of Laboratory E, Showing the Influence of Formal and Informal Organization

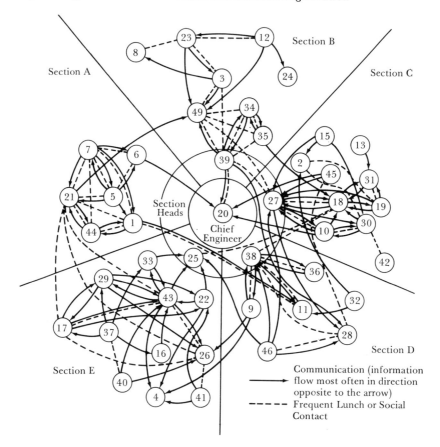

Figure 7.3 Communication Network in Another Department of Laboratory E, Showing the Influence of Formal and Informal Organization

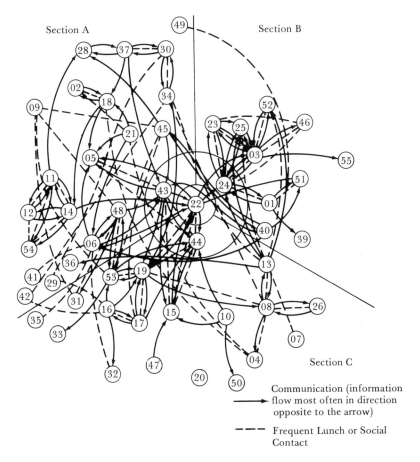

there is a higher degree of intersectional communication, and a high proportion of that occurs in conjunction with social contact. Causality cannot be inferred from simple correlations such as these, but there is some likelihood that the high degree of social contact among members of the three sections in figure 7.3 had something to do with their communication. Unfortunately, we do not know how either the social contacts or the communication contacts originally developed.

To more reliably test the relationships among communication, social contact, and formal organization, comparisons can be made of the degree to which overlap exists among them. A certain amount of overlap would be expected in any case by chance, so the amount of observed overlap can be compared to this chance level to see how much it exceeds a random level. The chance, or expected, level can be generated by means of a simple binomial model. To test the relation between sectional organization and the communication network, for example, each respondent in making his communication choices would be visualized as facing an urn with n_1 black balls and n_2 white balls. In this case, n_1 equals the number of people (other than the respondent himself) in his section; n_2 is the number outside the section. He makes m draws, where m is the number of people he chooses to communicate with. An expected distribution of white and black balls can be calculated for the m draws. The quantity m will vary with each individual since some people choose to communicate with more colleagues than do others. Summing the expected distributions provides a measure of expected degrees of overlap between communication choices and either formal or informal organizational relationships. The problem is very similar to the classical birthday problem (Feller, 1950) except that each individual is allowed to have several "birthdays" (the number of people chosen for communication), and the number of days in the year is set equal to the size of the total organization $(n_1 + n_2)$ from which the choices are made. The probabilities of overlaps expected at random in a twenty-person department for an individual choosing four persons for socialization and five for technical discussion are:

Number of choices common to both relations	0	1	2	3	4	5
Probability	0.27	0.47	0.23	0.04	0.001	0

Expected values for the total organization are obtained by summing all of the individual values. The observed distribution of overlaps is then compared with this expected distribution by means of a one-sample Kolmogorov-Smirnov test. The bar graph in figures 7.9, 7.10, and 7.11, shows the observed number of overlapping choices and the number predicted by the random model for laboratories A and B.

THE FORMAL ORGANIZATION

In both laboratories, the degree of overlap between the communication network, and formal organizational structure is significantly above chance (figure 7.4). This should not be surprising. One of the goals in creating an organization structure is to shorten communication paths. Those whose work is interdependent or who draw on similar kinds of expertise in performing their job are normally grouped together. The real goal of formal organization is the structuring of communication patterns. Knowing this, what more can be said about the impact of formal organization on communication? A lot. A number of different organizational structures have been used in R&D laboratories with very little real basis for choosing among them. A consideration of the way in which different organizational forms affect communication and the way in which communication, in turn, affects performance will, however, provide this needed basis.

The Trade-Off Between Project and Functional Organization

There are two conflicting goals that the structure of an R&D organization must meet:

1. The activities of the various disciplines and specialties must be coordinated in order to accomplish the goals of multidisciplinary projects.

Figure 7.4 Degree of Overlap Between the Technical Discussion Network and
the Formal Organization Structure in Two R&D Laboratories

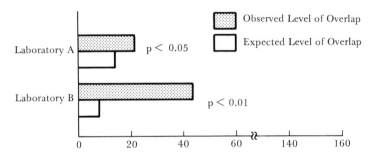

2. Projects must be provided with state-of-the-art information in
   the technologies they rely upon. This, as we have seen, is best
   accomplished through face-to-face communication.

It is the trade-off between these two goals that has resulted in the
various organizational forms used in R&D organizations. Functional management (figure 7.5), in which the laboratory is
organized around disciplines or technical specialties, best accomplishes the latter of the two goals. Project management (figure
7.6), in which all specialists assigned to a project report directly to
a single person, the project manager, and are usually moved into a
single physical location, best accomplishes the former.

Historically, the functional form of organization preceded all
others. In this type of organization disciplines or specialties are
grouped together, with one person (often a chief engineer or chief
scientist) in charge of each specialty. This is the way in which
most universities are organized, and it was only logical to extend it
to industrial R&D. This form of organization works well as long as
projects or tasks are primarily contained within single disciplines
or specialty areas. When systems projects require the collaborative
efforts of a number of different disciplines, this form of organization can result in serious coordination problems. No single individual is responsible for the project, and even when such an
individual is appointed, he confronts very serious problems in
coordinating and managing all of the subsystem interfaces when
the people responsible for subsystems are usually in different
departments reporting to separate bosses.

In response to the intraproject coordination problem and to the
need of certain types of customer for a single point of project
responsibility within the laboratory, project management came
into being. In this form of organization a single individual has
almost sole responsibility for management of a project. All
members of the project team report to him for work assignments,
and in many cases, he has authority to hire, fire, or transfer

Figure 7.5 A Typical Functional Form of Organization for an R&D Laboratory. The C_i are chief engineers or chief scientists responsible for disciplinary specialties within the laboratory. S indicates section or section head.

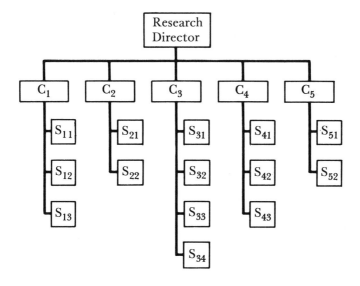

Figure 7.6 A Project Form of Organization Added to the Functional Organization. P_1 and P_2 are project managers.

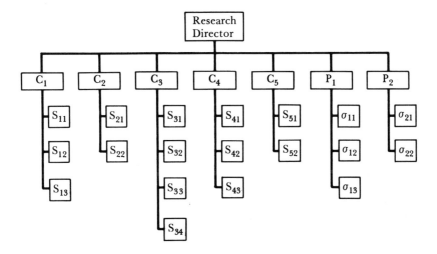

project members. In addition, those working on the project normally work at a single area near the project manager's own office.

Most other forms of R&D organizations are a variant or combination of these two forms. Over the years there have been many proponents of one or the other of the two. Hard evidence on their relative merits was sorely lacking until Marquis and Straight (1965) made a study comparing the two in terms of project performance.

In this study, data were gathered on thirty-eight fairly large (median size was $4 million), reasonably long-term (average duration was 3.4 years) projects. In addition to data on organizational structure, both technical performance and cost and schedule performance were also measured. The performance measures were obtained from responsible individuals in the government laboratories, which had contracted for the projects. A project was considered to be organized on a project basis if more than 50 percent of the assigned personnel reported directly to the project manager for work assignments, received merit reviews from him, and were physically located near him. If the percentage was less than this, the project was considered to be functionally organized. In all instances, a project falling on one side of the 50 percent point on one criterion did so on the other two as well. There were no ambiguous cases.

When they separated their data on the basis of organizational form and project performance, Marquis and Straight found quite different results for administrative personnel (contract lawyers, cost accountants, procurement officers, etc.) than they did for the technical personnel assigned to the project (Table 7.5).

In the case of administrative personnel, organization bears no relation to technical performance, but it does to cost and schedule performance. When administrative personnel were organized on a project basis, there was less likelihood of a cost or schedule overrun. Under project organization, the administrative personnel apparently were better able to control the activities of the project manager and technical personnel and avoid overruns.

Table 7.5 Two Measures of Success of Projects with Different Types of
Organization

	Administrative Personnel[a]		Professional Personnel[a]	
	Project (N = 17)	Functional (N = 18)	Project (N = 19)	Functional (N = 19)
Percent of projects rated highly successful technically	47	60	37	68
Percent of projects without cost or schedule overruns	47	22	39	34

Source: Marquis and Straight, 1965.
[a]N = Number of projects in each category. Three projects had no administrative personnel assigned.

In the case of technical personnel, there was no relation between
organizational form and cost and schedule performance. The
majority of projects overran their cost and schedule estimates
under both forms of organization. In contrast, technical performance was markedly better when technical personnel were
functionally organized.

This is at first a rather puzzling set of results. Why should cost
and schedule performance be better when administrative personnel
are organized on a project basis and technical performance be
better when technical personnel are functionally organized?
Perhaps a look at the nature of the work being performed will help
in understanding this phenomenon. The administrative personnel
are working in areas that are not changing rapidly. The state of the
art in cost accounting or contract law are certainly not as dynamic
as are most of the physical technologies employed in the thirty-
eight projects. For this reason, it is not as critical that these people
remain in close contact with their disciplinary colleagues. They
can afford to work full time on a project for three or four years,
largely out of touch with disciplinary colleagues and still not
become out of touch with the field. Therefore the second of the
two goals of organization structure loses its importance; the first
dominates, and project organization best accomplishes this first

goal. In the case of technical specialties this is not necessarily true. To be removed from developments in a dynamic technology for three or four years can have very drastic consequences. And since it should be clear now that the only effective way for most people to keep up with their field is through colleague interaction, the organization must be structured to promote this form of inter-action. Project organization does not promote this form of inter-action; indeed it impedes it. Functional organization does aid disciplinary interaction. Therefore better technical performance results under functional organization *in the case of long-term projects.*

Remember that Marquis and Straight's projects averaged 3.4 years in duration, and three or four years is sufficiently long to lose touch with many technologies. Had the projects been six months to two years in duration, their results might have been very different. This possibility now provides two parameters that must be used in the trade-off to determine the form of organization best suited for an R&D project. Technologies can be roughly ordered according to estimates of the rate of change of their respective states of knowledge (figure 7.7). This can be done subjectively, or indexes can be developed to aid in this estimation process. Certainly there should be agreement with respect to the relative positions of the physical and administrative technologies. And even among the physical technologies there should be some general agreement as to ordering. Structures technology, for example, probably lies toward the low end of the spectrum, while liquid crystal technology lies toward the more rapidly changing end of the spectrum. For a given project duration (T_1), the structures people could be moved to the project, while the liquid crystal people would be left in their functional department. For a longer project (T_2), both would be left in functional departments; for a shorter one (T_3), both would be shifted to the project. This results in a hybrid form of organization for the project, with the actual organizational location of people depending upon the nature of their tech-nologies and the duration of the project. The net result is a form

Figure 7.7 Organizational Structure as Function of Project Duration and Rate
of Change of the State of Knowledge in the Fields Involved in a Project

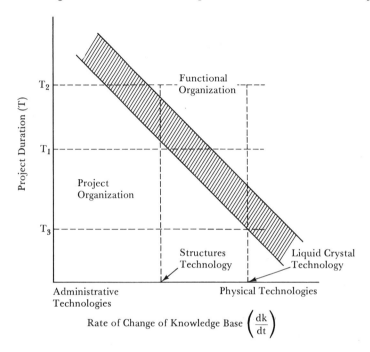

of matrix organization (figure 7.8) with more specific rules than are generally used for deciding the size and composition of the program manager's office.

The Matrix Organization The parameters of project duration and rate of change of the knowledge base in the required technologies can now be used to determine organizational form. There have been some rough rules for location of people in a matrix organization. The program manager's office generally contains some administrative and "systems level" personnel, the latter being responsible for resolving interface problems among the subsystems. We now have a more firm theoretical basis for organizational location. In the area to the left and below the diagonal in figure 7.7, a project form of organization is preferred. In this realm, the benefits of better internal coordination that are available under project management outweigh the benefits of improved disciplinary support that are available with functional organization. In the area above the diagonal, functional organization is preferable. Here, disciplinary support is needed to such an extent that it becomes reasonable to sacrifice some internal coordination for it. Certainly the administrative personnel belong in the program management office. In addition to them and to the systems people, however, many of the technical specialists can be located there as well. The decision will be based on estimates of the two parameters. This will allow better coordination within the project without unduly sacrificing connections to the technological bases of the project.

THE INFORMAL ORGANIZATION
The degree of overlap between technical discussion and social contact networks is quite marked in both laboratories (figure 7.9). There is very close agreement in the selection of individuals for social contact and technical discussion.

Even though it is impossible from data such as these to determine the direction of causality (that is, does social contact lead to

Figure 7.8 One Form of Matrix Organization. The P_i are program managers. Some of the people working on a specific project report directly to them, while others remain in their functional departments.

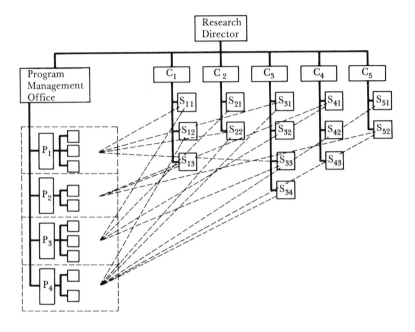

Figure 7.9 Degree of Overlap Between the Technical Discussion and Social
Contact Networks in Two R&D Laboratories

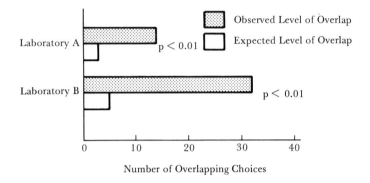

Number of Overlapping Choices

technical discussion, or do people socialize more with those with whom they like to discuss technical matters), we are led to conclude that the informal structure of the laboratory occupies an important position in the transfer of information. This, of course, lends further support for neutral social interaction as a mechanism to promote communication within the laboratory. Of course, the natural reaction to such a discovery is to question the degree to which it can be put to practical use. Management has no direct control over the informal organization. That is certainly true, but upon further examination, it becomes obvious that management already controls the informal organization, albeit through indirect means.

The question remains as to which of the two influences, formal or informal organization, is dominant and the degree to which they are independent of each other. To answer this question, we shall hold the effects of one of the two independent variables constant and measure the relation between the remaining variable and communication network structure.

A certain amount of overlap exists between measures of formal and informal organization (figure 7.10), but it is not significantly above what might be expected to occur by chance. In other words, the two can be assumed to be largely independent of each other. In reexamining the relationships between social contact and technical discussion while holding formal organization constant (figure 7.11), it can be seen that the relationship remains nearly as strong as before. The net conclusion is that although formal organization may be the more important of the two determinants of communication, informal organization makes its own independent contribution of nearly equal magnitude. This suggests a second and more serious question: given that informal organization is important to communication, what can management do about it? Management certainly cannot dictate friendships or force them to develop. On the other hand, it can create the necessary conditions. People must first meet in order to become acquainted. Management does, to a very great extent, control the processes by which people in an

Figure 7.10 Overlap Between the Social Contact Network and Work Group
Structure in Two R&D Laboratories

*Not statistically significant

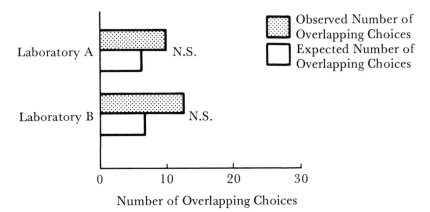

Number of Overlapping Choices

Figure 7.11 Overlap Between the Social Contact and Technical Discussion
Networks Outside the Work Group in Two R&D Laboratories

*Kolmogorov-Smirnov One-Sample Test

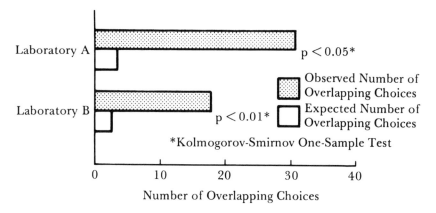

Number of Overlapping Choices

organization come to meet one another. Interdepartmental projects, which cut across existing organizational management, bring together people who might not otherwise meet. People who come to know one another through service on projects, proposal teams, task forces, and so on retain contact with the people whom they meet for some time even after the project or team has been disbanded. The project form of organization can and does provide indirect communication benefits over and above its immediate goal of coordinating the efforts of a diversity of disciplines. It has a lasting effect by way of increased interaction among its component disciplines, which persists beyond the termination of the project. For this reason, one can argue strongly for the use of project organization in those cases when the project duration is short or the technologies are relatively slow to change.

There are a number of devices available that allow management to influence communication patterns through the development of informal contacts (Helms, 1970). None of these guarantee that communication will result, but they all increase the likelihood of it. They range from short visits between persons who are organizationally or geographically separated through such temporary changes in organizational structure as project teams and task forces to more permanent changes such as reorganization and personnel transfer. The relative effectiveness of several of these devices was compared in an organization (Laboratory N) in terms of the number of communications that resulted in the year following the completion of the change (figure 7.12). It can be easily seen that the more enduring the change, the greater the effect. A transfer produces more communication than does membership on a project. A project produces more communication than a visit to another site, and so on. The cost of the more permanent changes is greater, however, often in both economic and human terms. This is difficult to quantify but should be taken into account in considering figure 7.11. The mechanisms that produce fewer communications may in fact be the more desirable ones from a total cost-effectiveness viewpoint.

Figure 7.12 The Relative Effect of Four Communication Multipliers on Inter-
location Communication in a Large Organization

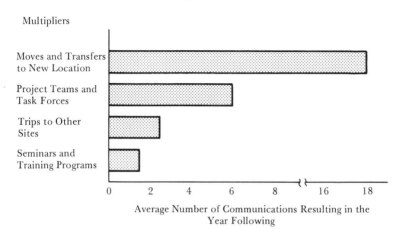

Average Number of Communications Resulting in the
Year Following

Before leaving the subject of communication multipliers, it should be pointed out that the data in figure 7.12 are inherently conservative. For example, they show only the communications reported by the transferred individual. The transferred person's effect extends far beyond this direct link, though. Probably his most important contribution lies in his ability to make referrals. The number of communication paths that potentially become available when a person is transferred is the product of the number of acquaintances that he developed in the two parts of the organization. For some people this can be a very large number, so it is possible that with only a very few transfers, a large number of communication paths can be created and coordination thereby improved.

Of course, the effect diminishes with time, since both people and activities will change in the old groups, and the transferred person will gradually lose touch. Kanno (1968) has shown that following a transfer between divisions of a large Japanese chemical firm, the transferred persons provided an effective communication link back to their old divisions for a year to a year and a half. The duration over which communications remain effective following a transfer is determined by many factors; principal among these are the rate of change of activities and turnover of personnel in the old organization. If projects are of short duration, with many new ones constantly being initiated, and the turnover of personnel is high, the effect of a transfer in promoting communication would be short-lived. Where the activity is more stable and turnover low, the transfer can be effective over a longer period of time. With estimates of these parameters and of the number of people with whose work the average [transfer] is acquainted, a systematic program of intraorganizational transfer can be developed. Such a program would contribute directly to communication, coordination, and empathy among the employees of the organization.

THE IMPACT OF STATUS ON COMMUNICATION

A number of studies (see, for example, Hurwitz et al., 1960, and

Newcomb, 1961) have shown that in the presence of prestige or status hierarchies, individuals of high status will tend to like and communicate frequently with one another and individuals of low status will neither like nor communicate with one another as much. In addition, lower-status members of the social system will direct most of their communication toward the higher-status clique, without complete reciprocation.

In Laboratory B about one-third of the members held doctorate degrees, creating an almost perfect situation for testing the effect of status on communication. Even a casual glance at figures 7.13 and 7.14 will show the impact of a status differential (in this case exemplified by possession of the doctorate) on the laboratory's communication network. Those with doctorates formed a tightly knit clique and apparently communicated quite freely among themselves, but they seldom socialized or discussed technical matters with the others. This lack of contact between the two groups could, of course, be quite disruptive to organizational performance, but an even more serious effect is evident. Those who did not hold doctorates rarely socialized with one another and discussed technical matters among themselves far less than did their colleagues with doctorates. Furthermore, they directed the majority of both their social contact (64 percent) and technical discussion choices (60 percent) to those who held doctorates, who in contrast directed only 6 percent of their socialization and 24 percent of their technical discussion choices to the other two-thirds of the laboratory's staff.

This phenomenon, observed many times before in controlled experimental situations but never in the field, has been explained by Kelley (1951) as a form of substitute upward mobility: "Communication serves as a substitute for real upward locomotion." Kelley goes on to qualify this statement by showing that it holds true only for those low-status persons who exhibit some desire to move upward. Cohen (1958) replicated Kelley's results with another experimental group and found further that one form of upward communication (conjecture about the nature of the

Figure 7.13 Social Contact Network in Laboratory B, Showing the Effect of Status on the Informal Organization (from Allen and Cohen, 1969)

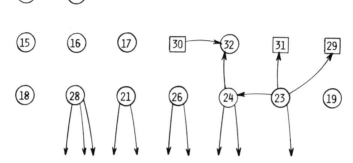

Graduate Chemists with No Ph.D.

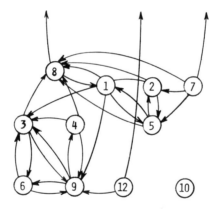

Ph.D. Chemists

☐   Nonrespondent

Figure 7.14 Technical Discussion Network in Laboratory B, Showing the Effect of Status on Communication (from Allen and Cohen, 1969)

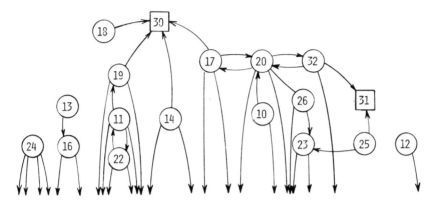

Graduate Chemists with No Ph.D.

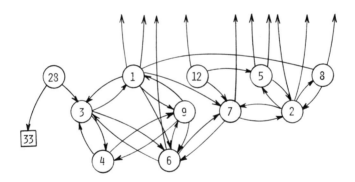

Ph.D. Chemists

☐ Nonrespondent

higher-status job) increases both when "locomotion is desired but not possible and where it is possible and where it is possible but not desirable."

The situation in Laboratory B can best be described as one in which upward mobility is highly desirable but, in the short run, impossible. It is therefore not surprising that the nondoctorates should attempt to enhance their own status through association with the higher-status doctorates in the laboratory. It can be argued that the nondoctorates receive their work assignments mainly from the doctorates and that they therefore turn to them for information. This is true to some extent, but it can explain neither the lack of downward communication from the doctorates nor the combination of a highly cohesive higher-status group coupled with a very noncohesive and even disintegrated lower-status group.

In an organization in which both doctorates and nondoctorates work together on the same projects, the most rewarding experiences, publication, recognition, and so on are almost bound to be restricted to those holding the advanced degree. This being the case, those without it are resigned to a strategy of gaining reflected glory as satellites of the higher-status group. They therefore tend to veer away from association with their lower-status colleagues and direct the majority of their attention toward those with advanced degrees, thus attempting to gain through association the upward locomotion that is in reality denied to them.

The problem can occur under other circumstances, of course. The possession of a doctorate is not the only determinant of status in a research laboratory. In addition to the obvious split between management and nonmanagement employees and between engineers and technicians, there can be status associated with the nature of the problems that people are working on. Thus a status differential can arise between engineers who are working on preliminary designs for future systems and those who are assigned to the development and testing of systems currently under contract. This can, and often does, create a serious disjuncture in the organization's communication system at one of its vital points. As another example, in university-affiliated laboratories, there can arise a serious status problem between faculty and nonfaculty scientists. This can cause serious problems of communication and undermine the effectiveness of the university affiliation.

The status problem raises its head in many places and many ways. There are no ready solutions, but the problem should be a serious concern to research managers.

NOTES

1. Expected frequencies are based on the total number of instances in which the source was referred to as a first, second, third, etc., choice.

2. It was interesting to note the development of certain social norms around the bridge games in this organization. In addition to being punctual and not keeping the players waiting, there is, for example, a taboo against technical discussions during the game. Such discussions are permitted during coffee breaks, however.

8 STRUCTURING ORGANIZATIONAL COMMUNICATION NETWORKS 2: THE INFLUENCE OF ARCHITECTURE ON COMMUNICATION

Previous chapters have provided evidence of the importance of communication among technologists, particularly with their own organizations. Some other factors have also been discussed that affect the structure or pattern of communications that develops in an organization. The present chapter will explore this subject further, with a consideration of the relative physical location of potential communication partners.

PROPINQUITY AND BEHAVIOR

There is a long history relating human interaction to relative location. Laboratory experiments by Leavitt (1951), Steinzor (1950), Sommers (1969), Strodtbeck and Hook (1961), and Hare and Bales (1963) have shown that not only does the explicit communication network affect the direction frequency of interaction in small groups but also that inherent in the seating arrangements is an implicit network that influences the interaction patterns of the group. For example, Steinzor found that within a circular seating arrangement, group members interacted more with individuals opposite them than with those adjacent to them. Sommers (1969) was able to increase significantly the incidence and duration of conversation among patients in a nursing home by rearranging the location of chairs in a dayroom. Sommers also documented in numerous settings the interaction between spatial factors and human interaction. A number of field experiments have also revealed the importance of propinquity in promoting interaction. Festinger, Schacter, and Back (1950) examined the relationship between physical distance and sociometric choice in a housing project and found that the most highly chosen individuals lived closest to their choosers; as distance from the selector was increased, the frequency of choice decreased continuously.

In an organizational context, Gullahorn (1952) studied the interaction pattern of twelve women in an office of a large eastern corporation. He found that distance was the most important factor in determining interaction, and when physical proximity did not account for interaction, friendship seemed to be the controlling factor.

Even the choice of a marriage partner has been shown to be significantly influenced by physical proximity (Abrams, 1943; Kennedy, 1943). Although it would seem that the results, in part, could be attributed to ethnic concentrations in neighborhoods, in 1940 35 percent of the marriages in New Haven occurred between individuals who lived within five blocks of each other.

In his study of a large educational institution, Maisonneuve (1952) discovered that propinquity was the variable that significantly influenced the formation of friendships among students in a classroom. He concluded that "very often people did not get closer to each other because they had a liking for each other, but they inclined to have a liking for each other because they are close to each other."

It would appear, then, that the physical layout may be a strong determinant of communication choices within an organization.

RESEARCH METHOD

In seven R&D laboratories, communication patterns were measured by either asking individual engineers and scientists to indicate which of their organizational colleagues they communicated with ("about technical or scientific matters") at a frequency of once a week or more or by sampling communication by means of a questionnaire administered weekly on randomly chosen days. The questionnaire listed the names of all professionals in the organization, and participants were asked at the end of a day to indicate the number of times they had communicated *about technical and scientific matters,* over the course of the day, with each colleague listed. Communications were sampled in this way for periods of from three to six months. A computation was then made to determine which pairs of individuals maintained regular communication at an average frequency of at least once per week.

The laboratories ranged in size from 48 to 170 professionals and included two laboratories in the aerospace industry, two in universities, one each in the chemical and computer industries, and one government agricultural research laboratory. Data were obtained from a total of 512 respondents in the seven organizations.

Distance between respondents was measured by means of facilities diagrams supplied by the organizations and by maps in the case of extended distances. In the first six cases, distance was measured from desk to desk; in the seventh organization distance was measured from building to building. In all cases, each engineer or scientist was taken, in turn, as a focal person and the actual walking distance from his desk to that of every other engineer or scientist in the organization was recorded. These were then aggregated in intervals of about ten feet, or three meters. In each three-meter interval (figure 8.1) the ratio was computed of the number of individuals with whom the focal person communicated to the total number of people available. Such a ratio can be computed for any frequency of communication (one or more times per month, one or more times per day, and so on); the present analysis is based upon an average frequency of one or more communications per week.

RESULTS

The ratio represents, on the average, the proportion of available people with whom an individual will communicate at a given distance and frequency. The aggregate ratios for all respondents are shown in figure 8.2 as the probability that two people will communicate about a scientific or technical subject matter. The scales of figure 8.2 are in logarithmic form in order to accommodate a range of separation distance from 2 meters to 255 kilometers. Figure 8.3 is a better illustration of the effect of distance on communication probability within the first 100 meters. Here it can be seen that the regression curve is a hyperbola with a correlation coefficient, $r = 0.84$.

In considering figure 8.3, the general shape of the curve is probably not surprising to anyone. One would expect probability of communication to decrease with distance. One might even expect it to decay at a more than linear rate. It is the actual rate of decay that is surprising. Probability of weekly communication

Figure 8.1 Method for Determining the Effect of Separation Distance on Communication (Frequency of Communication Held Constant)

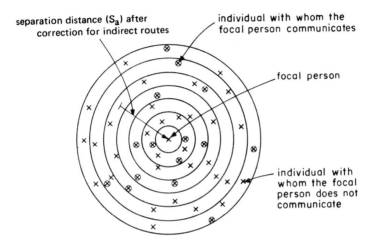

Figure 8.2 The Probability That Two People Will Communicate as a Function of the Distance Separating Them (100 Meters to 255 Kilometers)

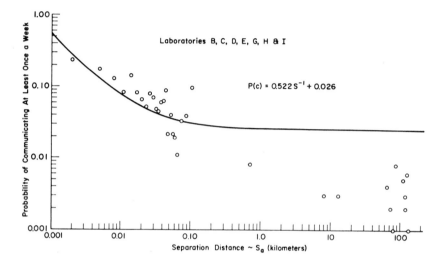

Figure 8.3 The Probability That Two People Will Communicate as a Function of the Distance Separating Them (0–100 meters)

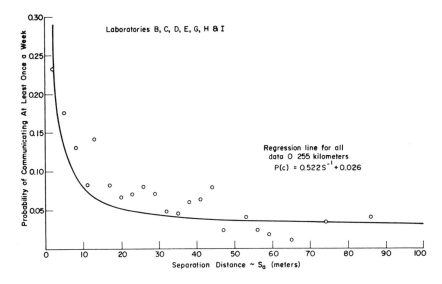

reaches a low asymptotic level within the first twenty-five or thirty meters. It is this extraordinary rate of decay more than the general shape of the curve that is so startling. For weekly contact, it is only within the first thirty meters that separation has any real effect on the probability of communication.

If communication frequency is increased or decreased, the curve of figure 8.3 shifts up or down accordingly, but the shape remains essentially the same. The frequency of once a week or more was chosen arbitrarily to represent regular and consistent communication.

ORGANIZATIONAL BONDS

One possible objection to figure 8.3 would be that any individual, for organizational reasons, is more likely to communicate with those nearest to him. Space in most organizations is allocated on a group basis, with people of similar background or people working on the same or similar tasks located near each other. Since an individual is more likely to communicate with those with whom he shares a common task or background, the decay in communication with distance is simply an artifact of organizational location. To test this possibility, the data from one of the seven laboratories were separated into two groups. Separate curves were then plotted for those pairs who shared an organizational group affiliation and those who did not (figure 8.4). The resulting curve for intragroup communication lies above that for intergroup communication as expected, but the general shape of the two curves is about the same as before. Common group affiliation merely shifts the relationship onto a higher curve, but the distance effect still operates in the same manner. The probability that an individual will travel a given distance to talk with someone in his group is slightly higher than the probability for someone in a different group. Both probabilities decay along a hyperbola and soon reach a low asymptotic level. The presence of the group bond merely introduces a relatively constant positive bias.

Figure 8.4 Probability of Communication as a Function of Distance—
Controlling for Organizational Structure

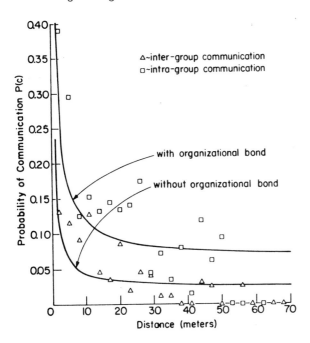

OFFICE ARRANGEMENTS

These results have a clear message for those who are responsible for the design of research laboratories. The classical approach of arraying offices in a linear fashion along a hallway (figure 8.5) maximizes the separation distance between occupants of the offices and is hardly the best way to promote communication.

To minimize separation, one should approach a circular or square configuration. Of course, in the case of large organizations it is impossible to have everyone within a thirty-meter radius. This would be true even if the building were circular. On the other hand, one need not go to the extremes seen in many research organizations, which look like an alphabet soup with buildings in the shape of H's, N's, Z's, and even W's. Usually such buildings result from a desire to give everyone an outside exposure. While it is very nice to be able to see outside of one's building and determine the weather, there are other ways to permit this without causing the extreme isolation that the elongated shape produces. One possibility is to move hallways to the exterior wall and provide windows along the hallways. Common areas, such as libraries, meeting rooms, or coffee lounges, can be given the windows. There are many ways of giving everyone some access to an outside view without arranging offices in such a way as to provide each one with such access. The less differentiation there is in the desirability of office locations, the greater the flexibility possible in making office assignments. One thing is certain: if the head of the organization wants to keep in close touch with what is going on in his organization, he must resist the temptation to locate his office in the corner with the best view. The center of the building is the place for him. This will minimize average separation between his office and the location of the groups reporting to him. Otherwise he is going to be further from some groups than others, with a corresponding degradation in communication.

VERTICAL SEPARATION

Vertical separation between floors in a building is another problem. The data would indicate that the vertical separation has at

Figure 8.5 The Classical or "Pigeon-Hole" Form of Office Layout

CORRIDOR

least as severe an effect as horizontal separation on communication (Peters, 1969). This effect depends upon many factors in addition to actual distance. The location of stairs or elevators, their accessibility (whether the stairs are protected by a fire door, for example), and the amount of visual contact that they allow all enter in. In most office buildings, the actual length of the stairs between two floors ranges between nine and fifteen meters, so it is reasonable to assume a one-story separation to be at least equivalent to that amount of horizontal separation. Elevators do not seem to change this situation. People are about as reluctant to use an elevator as they are to climb stairs. The difference between stairs and elevators becomes pronounced as the number of floors in a building increases. It can be assumed that while people's reluctance to travel on an elevator increases only slightly as the trip increases from one to ten stories, the difference between climbing one story and climbing ten on the stairs is considerable. As a matter of fact, people are probably much less than one-tenth as willing to climb ten stories as one. The data gathered thus far are incapable of either confirming or denying any of the above conjectures, but it nevertheless seems fairly clear that housing an organization in a tall building can lead to communication problems.

This does not mean that single-story buildings are desirable in all cases. After all, land values must be taken into account, and even ignoring this, as floor area increases, a point must be reached at which the average separation between offices in a single story building exceeds that in a multistory building of the same area.

An oversimplified example of this can be pursued by assuming a square building with no interior walls or corridors, each occupant assigned a ten-square-meter segment of the floor and with a staircase located in the center of the building. Floor area can then be increased and mean separation distance between occupants computed for cases in which that area is distributed over one or more floors. The length of the staircase will be assumed to be twelve meters.

Under these assumptions, there appear clear break points at
which mean separation distance will be decreased by adding floors
to the building (figure 8.6). The first three such points occur at
9,800, 17,000 and 23,000 meters, respectively. Moreover, the
break at 9,800 square meters involves a shift from one floor to
four floors. Given the initial assumptions of the problem, one
would never want to build a two- or three-story R&D laboratory.
The reason underlying this lies in the fact that in the region
between 9,800 and 17,000 square meters, the mean distance
between people located on *different* floors is always greater in a
two- or three-story building than in a four-story building with the
same floor space.[1]

Of course, the initial assumptions in the foregoing problem are
grossly oversimplified. For example, the effect of a staircase is
assumed to be the same as twelve meters of horizontal separation.
This accords with the data reasonably well for a single flight of
stairs, but it is probably not safe to assume that two flights are
equivalent to twenty-four meters, three flights to thirty-six meters,
and so on. The relationship is probably nonlinear, with the equivalent in horizontal distance increasing as the number of floors increases.

To see what happens when one tries to account for this, the
initial assumptions might be modified in the following way.
Assume again that the separation between the first and second
floors is 12 meters but that the distance between the first and
third floors is 2.25 times that distance, the three flights are equivalent to 3.75 times one flight, four flights equivalent to 5.50 flights,
and so on. The results are shown in figure 8.7. The first three
break points now occur at 9,000, 18,000, and 36,000 meters. Now
the first shift is from one to three floors, but one would still never
want a two-story building.

Once a laboratory building has more than two floors, one would
probably want to introduce elevators. Although people appear to
be just as reluctant to travel a single-floor separation by elevator as
they are to climb stairs, they are probably more willing to travel a

Figure 8.6 Number of Floors Required to Minimize Separation Distance
Between Occupants in a Building. This is a square building, with a staircase in
center, 10 square meters per occupant, and no internal walls.

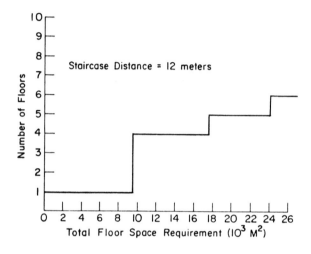

Figure 8.7 Number of Floors Required to Minimize Separation Distance Between Occupants in a Building. This is a square building, with a staircase in center, .10 square meters per occupant, and no internal walls.

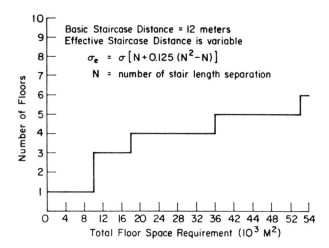

greater distance by elevator than by stairs. While the exact form of this relationship cannot be specified, at this time it is probably safe to assume that traveling ten stories in an elevator is not much worse than traveling one. This being the case, the most important break point in the above analysis is that at which a single floor stops being the most desirable solution. Since the optimal solution then involves more than two floors, an elevator would become desirable, communication would become less sensitive to building height, and land cost then becomes the sole governing variable.

The most important conclusion from this analysis is that, for communication purposes, a research manager should limit his laboratory to a single-story square building as long as the required floor space is less than 10,000 square meters. Above that area, the building should have at least three floors, and elevators should be installed.

INTERACTION-PROMOTING FACILITIES

The traffic patterns in any building certainly have a direct effect on communication. They both promote chance encounters and aid in the accomplishment of intended contacts. Much of the traffic in a building results from the movement of people to and from certain types of facilities they must use during the course of a day—among them, washrooms, copying machines, coffeepots, cafeterias, computer consoles, laboratories, special test equipment, supply rooms, and conference rooms. The types of facility that draw people to them will vary with the functions and operations of the organization. In all cases, they not only increase the occurrence of chance encounters among occupants of a building but often aid in promoting intended contacts by providing a person with more than one reason for traveling in a particular direction.

The presence of these interaction-promoting facilities should be taken into account when locating organizational groups within a building. It is seldom possible to locate everyone on the same floor within a thirty-meter circle; a possible way to counter undesired physical separation is to locate a specific facility (such as a wash-room or a laboratory) in such a way that it is shared by two groups whose physical separation might otherwise inhibit com-munication. The possibilities here are endless and require little imagination; only an awareness of the consequences of traffic patterns is needed to analyze any situation and locate and assign facilities and groups in ways that increase desired interaction.

AN EXPERIMENT IN LABORATORY DESIGN
In 1967 Laboratory G, the R&D department of a small chemical firm, was planning the construction of a new research facility, thus providing an opportunity to test some of the ideas presented here for improving communication. Discussions were held with the architects, and data from earlier studies were shown to them. The architects produced a design that minimized physical separation while providing for privacy and making use of laboratory space assignments to offset whatever office separations were necessary.

The building was laid out in a rectangular fashion, with offices in the middle and laboratory and pilot plant space at either end (figure 8.8). In the very center (room 503 in the figure) is a com-bination cafeteria and auditorium in which coffee is available, free of charge, all day. The purpose of this arrangement, of course, is to promote contact among those people housed in the various office clusters. Organizational groups are assigned to the small clusters of offices on either side (for example, rooms 617, 618, 619, and 620 are assigned to one group). Group size is generally in the neighborhood of four to eight engineers, chemists, and techni-cians, so a group can be accommodated easily within one or two office clusters. Managers are assigned to offices in the middle (515 through 518 and 525 through 528) so that the organizational bond between the manager and his group would guarantee com-munication. But then propinquity was necessary to insure com-munication among the managers themselves. Unfortunately, there are two clusters of managers' offices, separated by the lunchroom,

Figure 8.8 Configuration of the New Laboratory Facility for Laboratory G

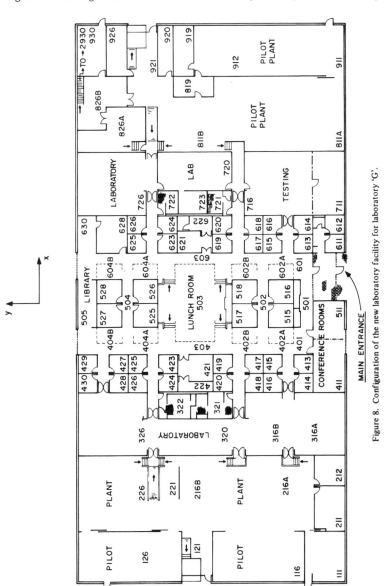

Figure 8. Configuration of the new laboratory facility for laboratory 'G'.

so not all managers can be kept in contact by office location. What was done was to use the location of the managers' offices to provide integration between groups separated along the x axis in figure 8.8 and then to provide for integration in the y direction by means of shared laboratory and pilot plant areas. The group occupying offices 413 through 420 shares a laboratory area with the group in offices 423 through 430. Their pilot plant areas are adjacent. The group in offices 623 through 626 shares a laboratory with the group in 617 through 620, and so on. To provide integration along the diagonals, interaction facilities—such as the lunchroom, coffeepot, computer consoles (room 421) and copying machines (room 621)—are all centrally located. The library (room 505) is also in a key position: it is the only part of the office area provided with windows. Perhaps an engineer who is checking the weather might check a current journal as well.

An additional feature of the building should provide more direct access to managers for their subordinates. Feeling that a secretary outside the boss's door might very often inhibit a subordinate from initiating informal contact with his boss, the architect located all secretaries around a corner out of sight of the manager's door (area 402, 404, 602, and 604). To provide communication between secretary and manager, a sliding window is provided on the secretary's side of the manager's office.

Laboratory G Before the Architectural Change
In order to determine whether the facility design accomplished its intended goals, communication measurements were made in Laboratory G both before and after moving into the new facility.

Prior to the opening of the new laboratory building, all employees at the main plant were located in several interconnected structures, most of which were originally build in the mid-nineteenth century to house a textile mill. Three R&D groups (molding materials, permeable materials, and fiberloys) were located with their laboratory space and pilot-plant facilities in the main plant building. Under this architectural arrangement offices were grouped into clusters with substantial distances between them and with routes between clusters that traversed production and inventory areas.

The approximate locations in the old buildings of the three major R&D groups are shown in figure 8.9. Offices of the molding materials and fiberloy groups were located on the same floor, roughly eighty-five meters apart. The permeable materials group was separated from the others by two floors. A forty-foot staircase contributed part of the 34 meters separating permeable materials from fiberloys, as well as part of the 104 meters separating permeable materials from molding materials.

The communication network for this organization in the old facility certainly reflects the spatial arrangement (figure 8.10). The three principal groups are not in very close communication, with molding materials especially showing the effect of its isolation.

Since the three groups are of roughly comparable size, we can compute an index of the level of communication between each pair as follows:

$$C_{AB} = \frac{\displaystyle\sum_{i=1}^{N_A}\sum_{j=1}^{N_B} C_{ij}}{N_A N_B}.$$

where:

C_{AB} = strength of communication bond between groups A and B
C_{ij} = 1, when person i (in group A) reports weekly communication with person j (in group B) or vice versa
. = 0, otherwise
N_A, N_B = number of professionals in groups A and B, respectively.

The strengths of the communication bonds among the three groups are 0, 0.11, and 0.46, respectively. Direct communication was high between only two of the groups.

An index of intragroup communication can be constructed in a similar fashion:

Figure 8.9 Approximate Location of R&D groups. This shows the office and lab space at Laboratory G in the old buildings.

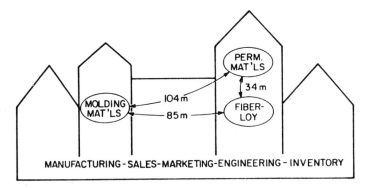

Figure 8.10 Laboratory G Prior to Architectural Change (One or More Communications per Week)

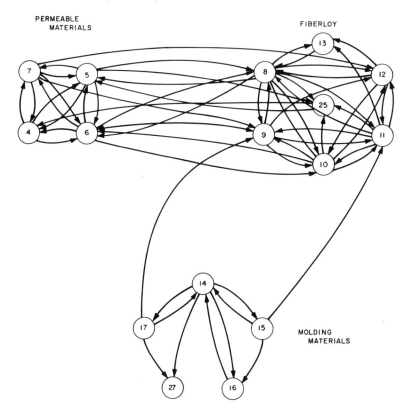

$$C_A = \dfrac{\displaystyle\sum_{k=1}^{N_A}\sum_{i=1}^{N_A} C_{ki}}{N_A\,(N_A - 1)}$$

where:

C_A = strength of communication index within group A

C_{ki} = 1, when person k reports weekly communication with person 1 or vice versa

. = 0, otherwise

N_A = number of professionals in group A.

The strength of communication index varies considerably for the three groups, from a low of 0.15 to a high of 0.67. This can be attributed, at least in part, to the arrangement of offices in the old building where molding materials was less dispersed than the other two groups.

Laboratory G After the Architectural Change

Following the opening of the new building, the three R&D groups plus one new group (formed partially out of fiberloy) were located approximately as shown in figure 8.11. The distances shown in the figure are mean distances between offices in which the location of the group managers' offices (although separated from the rest of the group) are included in the computation.

The most significant effect of the new facility is a reduction in distances between groups (table 8.1). The mean reduction in inter-

Table 8.1 Distances Between R&D Groups

Separation between	and	Distance in Meters	
		Before	After
Molding materials	Permeable materials	104	15
	Fiberloy	85	22
	Printing materials		23
Permeable materials	Fiberloy	34	22
	Printing materials		19
Fiberloy	Printing materials		14

Figure 8.11 Location of R&D Groups in the New Building

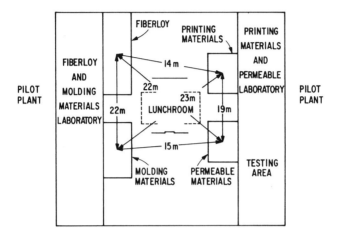

group distance is about 73 percent. This by itself should certainly improve communication among the groups, and, in fact, it does (figure 8.12). Unfortunately, there is now the additional complication of a new group (printing materials), plus transfers and turnover among the staff, but there can be little disagreement that intergroup communication has increased. The number of people having weekly contact increased substantially for two of the group pairs and remained essentially constant for the third (table 8.2). Another way of looking at this is to consider the number of communications between groups over a given time period (table 8.3). In all three cases, there is an increase in the number of communications per potential communication pairs per week. Communication among the three original groups is markedly better in the new building. At the time that the second set of measurements was made, there appeared to be a serious problem with the new R&D group, printing materials. This group was formed by moving several people from the fiberloy group, along with people moved in from other locations and some new employees. Communication between printing materials and fiberloy is relatively strong as a consequence. There is virtually no communication between printing materials and either of the other two groups. This situation may correct itself with time, and the new facility may be helpful in accomplishing this.

Table 8.2 Communication Bonds Among R&D Groups (Based on an Average of One or More Communications per Potential Pair per Week)

Communication between	and	C_{AB} Before	After
Molding materials	Permeable materials	0	0.47
	Fiberloy	0.11	0.36
	Printing materials		0.04
Permeable materials	Fiberloy	0.46	0.45
	Printing materials		0
Fiberloy	Printing materials		0.36

Figure 8.12 Laboratory G Following Architectural Changes (One or More Communications per Week)

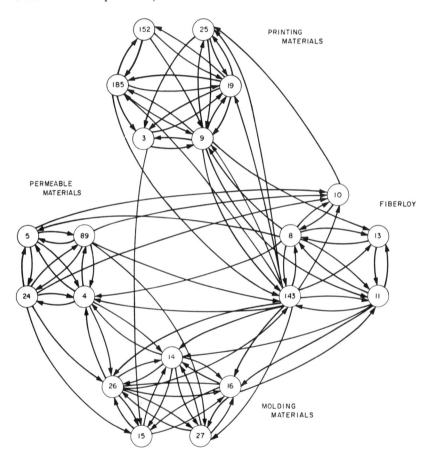

Table 8.3 Number of Communications Between R&D Groups

Communication between	and	Communications per Potential Pair per Week	
		Before	After
Molding materials	Permeable materials	0.35	2.89
	Fiberloy	1.74	2.49
	Printing materials		0.75
Permeable materials	Fiberloy	2.77	3.81
	Printing materials		0.23
Fiberloy	Printing materials		4.89

Communication within each group should also be affected by the move into the new facility. Offices are now grouped together more closely, and the layout is specifically designed to promote intragroup communication. Only one of the three groups had what could be called reasonably good communication in the old facility. Following the move into the new building, there is a marked improvement in intragroup communication for all three groups (table 8.4).

Some additional comments should be made concerning the permeable materials-fiberloy link. There is some explanation for lack of improvement in that case. This link was altered not only architecturally but also through the imposition of a "people barrier," the new printing development group having been placed between the fiberloy and the permeable offices. This barrier is particularly effective because the fiberloy and permeable laboratories are near their respective office areas and at opposite ends of

Table 8.4 Communication Level Within R&D Groups (Based on One or More Communications per Pair per Day)

Group	C_A	
	Before	After
Molding materials	0.15	0.30
Permeable materials	0.67	1.00
Fiberloy	0.17	0.45
Printing materials		0.37

the building. Thus, the two groups are not even forced to cross the barrier. This is shown in figure 8.13 in terms of a before and after comparison. The differences in the postmove arrangement may more than offset the advantage of closer distance.

The Effect of Interaction Facilities
In this case, the interaction facilities to be examined are chemical laboratories. Each of the four groups was assigned a laboratory area, two groups to each side of the building. Printing materials and permeable materials share a laboratory, as do molding materials and fiberloy. Contrary to prediction, however, the sharing of a laboratory area did not promote intergroup communication (table 8.5). As a matter of fact, those organizations with both shared laboratories and adjacent offices have the weakest average communication bond. Referring back to figure 8.8, communication occurs most easily in the x direction. This could be the result of another "facility," the manager's office, which tends to draw people in that direction. It appears at this juncture that shared laboratory space does not promote much communication.

Table 8.5 Influence of Shared Laboratory Space on Communication

	Office Areas Adjacent, Labs Separate	Labs Adjacent, Office Areas Separate	Labs and Office Areas Adjacent	Labs and Office Areas Separate
Mean distance between offices (meters)	13.5		19	20.5
Mean communication bond (C_{AB})	0.42		0.18	0.25

The Effect of Distance Reduction
There can be little doubt that communication among the three R&D groups improved in the new facility. A question remains whether it improved in a predictable manner and whether from this study one could in turn predict the effect of future architec-

Figure 8.13 Before and After Comparison of the Situation Between Permeable Materials and Fiberloy

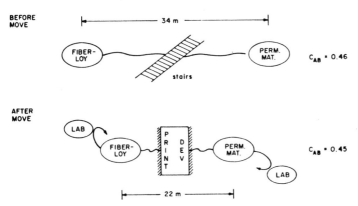

tural changes. The changes in distance and communication can be reduced to common terms by two measures. The first is the ratio of the before and after distances: D_1/D_2.
where

D_1 = distance between any two groups in the old facility
D_2 = distance between the same two groups in the new facility
and the second is the ratio relating the communication indexes:

$$\frac{C_{AB2} - C_{AB1}}{1 - C_{AB1}}$$

where

C_{AB1} = communication index between any two groups in the old facility
C_{AB2} = communication index between the same two groups in the new facility.

The distance ratio is simply a measure of relative distance change and can theoretically vary from zero to infinity. The second ratio is the relative increase in the completeness of the intergroup connection.[2]

When the three values of $\frac{C_{AB2} - C_{AB1}}{1 - C_{AB1}}$ are plotted as a function of $\log D_1/D_2$ (figure 8.14) they fall close enough to a straight line passing through the point (1,0) to at least arouse curiosity.[3]

Communication with Other Departments

In addition to the communication changes that are internal to the laboratory (the changes in linkage between R&D departments), there were also external communication changes. Recall that prior to the move, the various R&D groups were located in the midst of the other organizational components (figure 8.9) and afterward they were separated from them by the move into the new research center. Although the actual distances from non-R&D groups were never measured, it is clear that they all increased as a result of the move. A comparison of the communication indexes before and after the move between the R&D laboratory as a whole and each

Figure 8.14 Change in Communication Bond as a Function of Change in Distance

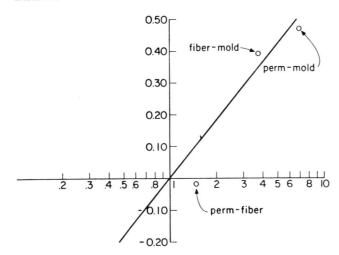

Figure 8.15 Communication with Other Parts of the Firm

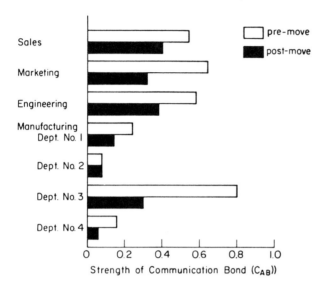

of the other organizational groups can be examined to indicate the general effect of the separation. The data are shown in figure 8.15. In all cases but two, the communication bond decreased in the postmove period. This suggests that the move to the new laboratory generally diminished R&D's communication bonds with the other parts of the organization.

The changes that were introduced to improve communication in the R&D laboratory had the inadvertent side effect of reducing communication with other parts of the firm. There are a number of remedies that can be proposed here. The general idea is to create reasons for the movement of people between the R&D laboratory and the other departmental areas. The firm is presently experimenting with some of these. The problem does not appear insurmountable, and external communication should be returned to its old levels without relinquishing any of the gains in internal communication.

NOTES

1. See Fusfeld and Allen (1974) for a more complete explanation of this phenomenon.

2. It represents the change in communication $(C_{AB2} - C_{AB1})$ relative to the potential for improvement $(1 - C_{AB1})$. Hence, if $(1 - C_{AB1})$ is regarded as the unfilled communication bond, then $(C_{AB2} - C_{AB1})$ is the relative increase in the completeness of the connection. Of course, there is only *real* increase in the completeness of the network when the ratio is positive, the term being phrased in the positive sense to indicate a measure of improvement in the postmove condition. In addition, the term is designed to normalize any change in C_{AB} relative to the potential and to theoretically vary from negative infinity to 1.0.

3. If the change in communication were to result from changes in individuals' perception of distance, then the relation between relative distance and communication levels might be expected to follow the laws of psychophysical scaling. A number of investigators (Vincent et al., 1968; Kunnapas, 1958; Gilinsky, 1951) have in fact shown subjects' estimates of distance to be a power function of real distance. While three points are hardly enough to build a strong case, it is interesting that they should fall so close to a function of the same general form.

9 A FIELD EXPERIMENT TO IMPROVE COMMUNICATION: THE NONTERRITORIAL OFFICE

As serious as the effects of physical separation on communication may have appeared in chapter 8, we have yet to take into account further effect of indirect routing. In analyzing communication it gradually appeared that in some cases the effect of distance was greater than in others. Looking a bit further, it seemed that the degree to which the route from one person's location to another's was direct or indirect seemed to have something to do with this.

THE EFFECT OF INDIRECT TRAVEL ROUTES ON COMMUNICATION PROBABILITY

One further aspect must be considered before leaving the subject of architecture. As though physical separation were not serious enough, there appear to be circumstances that can exacerbate its effect. The amount of difficulty, by way of corners to be turned, indirect paths to be followed, and other obstacles encountered in traversing a path intensifies the effect of separation on communication probability. One index of this difficulty, something that might be called a "nuisance factor," is the difference between the straight line and actual travel distances (figure 9.1) separating two people. When communication probability is plotted as a function of the magnitude of the "nuisance factor" (figure 9.2) the effect is quite startling. This holds true whether the nuisance factor is computed on an absolute basis or as a proportion of straight line separation distance.

These data, naturally, argue for the removal of as many walls as possible. In effect, they provide strong support for the old-fashioned open-bay arrangement. This should not be startling to anyone who has worked in an open bay. It is just marvelous for communication. The problems it causes all involve noise or excessive communication, which can be controlled. Carpeted floors, which absorb noise and are competitive in cost with tiled floors, are essential. Some office landscaping schemes provide insulation from both visual and aural noise, while allowing the free flow of people and communication. Certainly bays should not be too large. The maximum size is difficult to state exactly, but the

Figure 9.1 A Comparison of Straight Line (S_L) and Actual Walking Distance (S_A) Between Two Offices

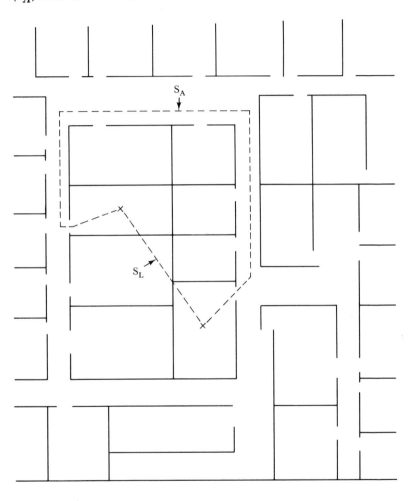

Figure 9.2 Probability of Communication as a Function of Magnitude of the Nuisance Factor

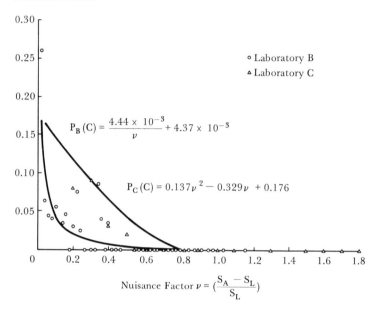

$$P_B(C) = \frac{4.44 \times 10^{-3}}{\nu} + 4.37 \times 10^{-3}$$

$$P_C(C) = 0.137\nu^2 - 0.329\nu + 0.176$$

Nuisance Factor $\nu = \left(\dfrac{S_A - S_L}{S_L}\right)$

extremes that were common in aircraft industry of the 1950s are unnecessary. Even bays need not be totally enclosed. Partial walls, which extend from floor to ceiling but stop considerably short of totally enclosing an area, should be effective in overcoming the feeling of vastness in a bay. They should screen some of the noise, as well. There are many things of a very imaginative sort that can be developed following from the data in this chapter.

THE NONTERRITORIAL OFFICE EXPERIMENT

The data and discussion to this point have emphasized the impact of physical layout on the communication of an entire organization. The effect at the small department or group level can be very great too.

Interpersonal communication patterns that evolve among those occupying a particular office area, laboratory, building, or other division are especially susceptible to architectural constraints. A prime determinant of communicator choice is the physical distance separating the parties in the organization. Opportunity for establishing eye contact with potential discussion partners and the sharing of equipment or physical space are important for developing personal contacts. These contacts are the prime vehicle for transmitting ideas, concepts, and other information necessary for ensuring effective work performance.

The more diverse the training and experience of a group's personnel, the more it can benefit from an open exchange of problems and ideas among its members. In this manner, a group can achieve greater problem-solving effectiveness. Where shared information will enhance the quality of group output, isolating individuals from their colleagues must be avoided. This is not surprising once one recognizes that employees are the principal repositories and disseminators of an organization's expertise. It is primarily through personal contacts with organizational colleagues that an employee, particularly a newly hired one, gains access to the wealth of experience that the organization possesses.

Product Engineers

The need for information exchange is particularly acute among
product engineers, who play a special role in the organization.
They mediate between the R&D and production departments and
assume responsibility for maintaining product quality from the
initial point in the production process through its eventual use in
the field. There is no organized body of literature to which the
product engineer can turn when faced with a new problem.
Instead he must rely on his own experience or on the experience
of others. At such times good communication becomes essential,
for it is only through good communication that knowledge of his
problem can reach a colleague with relevant experience. Interpersonal communication provides the essential link between a
problem and the experience required to solve it. Improved
communication within a product engineering department should
lead to a sharing of problems and a sharing of information and
experience essential to their solution.[1]

The Experimental Department

The product engineers in this study were all members of a single
department. Department size varied from thirteen to nineteen
members over the course of a year's study; altogether data were
gathered from a total of twenty-four individuals. At the beginning
of the study, the department was housed in a fairly standard arrangement with either one or two persons assigned to each of
several offices, which were strung along a corridor. The department also maintained laboratory facilities located directly adjacent
to its office area (figure 9.3).

The Nonterritorial Office

The nonterritorial office was designed specifically to improve and
increase the sharing of problems and experience.[2] It is an "open
floor plan" type of office, but goes far beyond any of the open
floor plans or landscaped offices. Under this concept, not only are
all office walls removed, but most desks and other permanent

Figure 9.3 Original Floor Plan of the Department

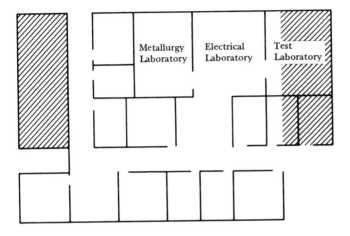

stations are eliminated as well. There remains but one permanent station, occupied by a "central communicator" who handles incoming and outgoing mail, assists visitors, and operates a switchboard directing calls to the telephone nearest the recipient of a call. All work is performed at laboratory benches and large round tables, and an individual may choose to work anywhere that suits him in the area or that is convenient. In the experimental department, an electronic components laboratory remains in its former position, adjacent to the old office area, but it is no longer enclosed. There is now free access between the table area and the laboratory area. In addition to the laboratory, there are three other special areas (figure 9.4). A computation area is partially screened to contain noise and houses consoles for access to a computer. A quiet area, enclosed by one wall and a drapery and containing comfortable chairs, can be used for meetings, performance evaluations, or work requiring high concentration and minimum disturbance. Finally, a total quiet room (formerly the department head's office) has been retained so that an individual or group can work behind a closed door.

The area is attractively and tastefully decorated and a number of items, such as carpeting and cloth murals, have been provided to reduce noise level.

The Research Methods Used

All measurements, with the exception of the performance measures, were applied to both the experimental department and a control department doing similar work at another of the company's plants about two hundred miles away.

Since the principal effect expected is an increase in communication, communication was measured at three levels: within the experimental department, between the experimental department and other departments at Essex Junction, and outside the Essex Junction plant. All of these were measured by a single-page questionnaire administered weekly on random days for a period of three months prior to the introduction of the nonterritorial office

Figure 9.4 Floor Plan of the Nonterritorial Office

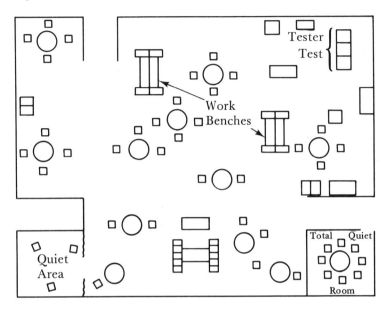

and eight months after the facility change. The questionnaire listed the names of all those in the department and required only that a number be circled to report the number of communications with someone on the sampling day. Since two reports were therefore available on each communication, a simple reliability check could be made on the basis of unreciprocated reports. Communications outside the department were reported in a similar way, with the single exception that names had to be entered by the respondent.

Removal of office walls, even without the removal of desks and permanent stations, can often be very upsetting to employees, so employee acceptance of the new scheme had to be ascertained. This was done by a questionnaire administered two months prior to the change, three months after the change, and again five months after that.

Since there was a distinct possibility that the nonterritorial concept would fail because occupants might still "stake a claim" to their own territories within the open area, a measurement had to be taken of the choice of seating position. This was done with a large diagram showing the locations of all tables and work benches. On the diagram, an assistant wrote the names of people in approximately the location they were sitting or standing at the time. This was done on the communication sampling days at 11:30 A.M., 2:30 P.M., and 4:00 P.M.

Performance was measured by means of interviews with members of departments that had often served as "customers" of the experimental department. Key personnel in the customer departments were asked to rate the experimental department in terms of four job dimensions.

Unless otherwise specified in the sections that follow, only measurements obtained from those individuals who remained in the experimental department throughout the study period will be reported. This is one of the serious limits imposed on longitudinal field research. Over the period of a year, internal turnover of personnel reduced the number of individuals participating in the study from nineteen to thirteen. Two more engineers were later transferred out of the department (three and five months after conversion), but they had both submitted a sufficient number of pre- and post-change communication questionnaires to be included in the comparison. One other person was transferred shortly before the change but returned four months before completion of the study. His communication patterns were therefore included in the comparison. Four other engineers were transferred into the department after the change. Their data will be shown only in those instances in which they shed additional light on some issue.

Employee Satisfaction with the Nonterritorial Office
In the opinion of the investigators, employee satisfaction with the nonterritorial office was the critical ingredient in the whole study. Could people adapt to the idea of working without a personal home base? Much has been written in recent years about the instinctive drive to claim and defend a personal territory. While hardly subscribing to all the claims for a "territorial imperative" among humans, we remained skeptical when it came down to removing all vestiges of personal space from a person's working environment. The amount and type of personal space has become one of the principal means of communicating one's status in an organization, and the opportunity to decorate a personal space has become one of the few remaining avenues for expression of individuality in large organizations. The removal of both of these, it seems, would almost surely produce dissatisfaction. This can be clearly seen in the amount of fear that is aroused in most people when presented with the possibility of having to work in a nonterritorial office.

For these reasons, it was essential that employee satisfaction with the arrangement be measured along as many dimensions as possible. Measurements were made in both the experimental and control groups two months before the change and eight months after the change. As the result of internal personnel transfers, there were only ten people in the experimental department from whom a valid before and after comparison of satisfaction could be made.

In general, the feelings among department members about non-territoriality had shifted in the favorable direction (figure 9.5). It appears that the longer a person works under this arrangement, the more favorably disposed he becomes toward it. There was a fairly even range of responses in the before measurement, with five out of ten indicating a negative or, at best, an indifferent attitude toward it. After eight months, one person remained indifferent; all the rest showed positive responses.

Although the actual amount of space had not changed, people generally felt as though there were more space (table 9.1). Even more surprising (although neither shift is statistically significant), occupants felt that they had more privacy in the nonterritorial office, while at the same time reporting the amount of distraction to be slightly higher. These results seem less contradictory upon actual observation of the behavior of occupants. In the non-territorial office, it is actually easy to bury oneself in a corner and avoid distraction. If someone is sitting in certain places, it is obvious that he wants to be left alone. Norms seem to have developed around this, which allow a person to control his privacy and the amount of distraction he confronted.

Table 9.1 Mean Level of Satisfaction with Work Environment Before and After Introduction of the Nonterritorial Office (5 point scale: 5 = high, 1 = low)

	Experimental Department			Control Department		
	Before	After	p^a	Before	After	p^a
Amount of space for the job	3.33	3.78	0.05	3.60	3.92	NS
Amount of privacy	3.42	3.78	NS	3.87	3.73	NS
Amount of noise	2.69	2.50	NS	3.96	3.83	NS
Amount of distraction	2.96	3.24	NS	3.50	3.40	NS
Ease of communication	2.14	3.33	0.005	3.33	3.73	NS
Feeling about working in nonterritorial office	3.23	4.19	0.05			

[a]Wilcoxon Matched-Pairs Signed-Ranks Test

Figure 9.5 Comparison of Feelings about Working in the Nonterritorial Office
Before and After the Experience

— — — June 1970

———— April 1971

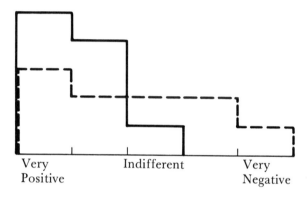

Very   Indifferent   Very
Positive   Negative

One of the more prominent fears in moving to any form of open bay configuration is that of noise level. In this case, the perceived level of noise increased, but not enough to be considered significant statistically. Considerable effort had been exerted to prevent the level of noise from rising too much. Carpeting was installed over the old floor tiles, and fabric murals were used to decorate the walls and help to dampen the sound. Neither of these features was exceptionally costly, but they indicate the care that was taken to prevent the noise problem from occurring. One important aspect of the work should be noted: there was very little need for typing by the members of the experimental department. Any required was performed by a typing pool located some distance away. Typewriters and telephones can be annoying sources of noise in an open plan office and care must be taken in dealing with them. In this case, all incoming telephone calls came to a single switchboard. The switchboard, of course, buzzed until someone answered it: but because it was at one location, the noise was somewhat isolated and the delay in answering was usually short. When an incoming call was redirected to an extension, a single buzz usually sufficed to draw the attention of the intended recipient.

In another instance of an open-plan office, which the author observed, engineers and scientists were located in an open bay, while all typing and answering of incoming calls took place within closed offices located along one wall. Incoming calls were redirected to the proper extension and a light was flashed on the extension phone to attract the attention of the person for whom the call was intended. There are many possible ways to cope with the noise problem in open areas. The specific nature of the work being performed and local circumstances will dictate different solutions in each case. The main point to be made here is that the noise problem is not an intractable one.

Returning to the situation in the experimental department, it seems safe to say that direct exposure to the nonterritorial office reduced the fear that seems inherent in the idea. The longer people worked in this type of environment, the more they liked it. In addition to the questionnaire data, several of the engineers volunteered their opinions to one of the investigators, with such comments as, "Don't ever fence me in again!" and "I was skeptical before, but I'd hate to go back to closed offices now." It certainly seems that our apprehensions about employee acceptance were laid to rest. Of course, we might have nothing here but a Hawthorne Effect, but it seems doubtful that such a condition would persist for over eight months with relatively sophisticated workers. Furthermore, if positive responses could be prompted simply as a result of the special treatment accorded the department, there would not have been so many negative responses in June. The "special treatment" had actually begun some time prior to June at the time when the department was selected for the experiment and first told about it. All members of the department had viewed scale models of the new facility arrangement for several months before the June survey and had seen or talked with the architects and designers, who almost constantly visited the area during the spring of 1970. If being specially selected elicits positive feelings, then those feelings should have been evident at the time of the June survey. The control group, which was told that they had been selected for a study on which future facilities planning would be based, did not shift significantly on any of the dimensions of satisfaction (table 9.1).

INTRADEPARTMENTAL COMMUNICATION PATTERNS

Among the members of the experimental department, communication increased significantly[3] both in terms of the number of communications per person ($p < 0.02$) and in the number of individuals ($p < 0.01$) with whom the average engineer communicated (figure 9.6). In the original office arrangement, the pattern of communications was very strongly influenced by the positioning of offices. An individual communicated a great deal with his office partner and perhaps with a next-door neighbor but there was little tendency to go much further. As expected, this disappeared.

Figure 9.6 Communication Among Department Members Before and After
the Introduction of the Nonterritorial Office

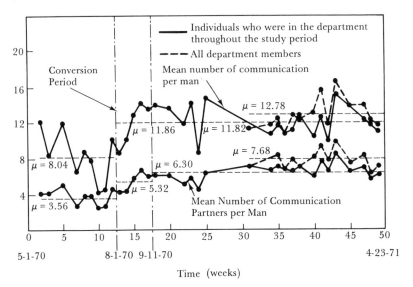

Time (weeks)

Communication was now more evenly spread through the department. It was also far heavier. Prior to the facility change, the average engineer communicated with a department colleague 8.04 times per day, or about once an hour. These communications were held with 3.56 different individuals for an average of 2.26 communications per person. Following the change, the number of daily communications increased to 11.82, and communications were held with an average of 6.30 individuals. This results in a rate of 1.88 communications per individual. In other words, while the number of daily communications increased under the new scheme, the number of people with whom an individual communicated increased at an even greater rate. The average engineer, therefore, had daily contact with a higher proportion of the members of his department under the new scheme.

It is important to note that the number of people with whom an engineer communicated actually increased over a period in which the pool of available communication partners was shrinking. At the outset of the study, there were nineteen people in the department. Over the course of the study this number gradually shrank to a low of thirteen; only in the closing months were new members introduced and one former member returned, bringing department size back up to sixteen. When new members are taken into account (dashed line in figure 9.7), the average number of people with whom communication was held increased to 7.68. This is more than double the initial level.

THE TERRITORIAL IMPERATIVE REVISITED

Basic to the concept of the nonterritorial office is the implicit assumption that people will not remain at the same work station but will position themselves wherever they can work most effectively at a given time. If people "stake out" their own territories and remain within them, the facility becomes no different from any other open-plan office. With no previous direct experience upon which to draw, the question of the occupants' reaction remained an important one until the time of the experiment.

Figure 9.7 Proportion of Time Spent Working at Tables That Is Allocated to
One Specific Table

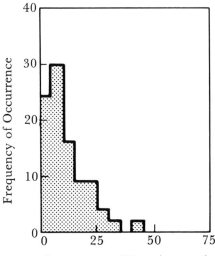

Proportion of Time (percent)

There was no way to determine whether individuals would decide
upon favored positions and then spend most of their time at those
stations or vary their location from day to day and even within
each day. The limited evidence available at the time of the experiment tended to favor a tendency to establish personal territory.
Studies in nursing homes and mental hospitals (Sommers, 1969)
indicate that the occupants of such institutions frequently establish personal territories, whether a particular seat at the dinner
table, a favored chair in a lounge, or even a specific tree to sit
under on a summer afternoon. Furthermore, they can become
upset when someone preempts what they consider their special
territory. Students tend to favor the same seats in a classroom over
a term. Even at home, most of us have a favorite chair and will be
quick to assert our territorial rights should it be invaded by one of
the children or sometimes even by a guest.

From this point of view, the chances for successful operation of
the nonterritorial scheme appeared slim. To help offset this, occupants were advised that they could keep no personal artifacts in
the new area. All photographs and even books had to be taken
home. Needed personal books would be replaced by the company
and remain departmental property. While this approach seemed
necessary, it was feared that it might engender some resentment
on the part of participants. To see whether it could be the source
of any dissatisfaction indicated later, an inventory was made of
the number of personal artifacts displayed by each engineer in his
former office. These ranged from a single motorcycle helmet to
several family photographs and a series of company awards and
engineering certificates. The inventory also included maps, plants,
office equipment, and drawing easels.

In fact, rather than laying claim to any specific position, the
occupants seem to prefer to move about considerably over the
course of a day. No one spent more than 50 percent of his at-table
time at a single table, and the median proportion of time allocated
to a single table by an individual is less than 10 percent (figure
9.8). People do have preferred tables, but there are usually two or

Figure 9.8 Proportion of Time That Various Proportions of the Total Departmental Complement Are Present in the Office Area

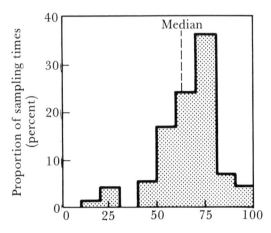

three of these and they tend to be in very different parts of the office area. An engineer typically can be found moving back and forth between these tables during the course of a day. Specific tables also become identified somewhat with function. Those near the laboratory benches are used when considering or discussing test results. Tables near the windows seem to be used more for solitary, analytical work. The total quiet room and partial quiet area were seldom used. In seventy-one samplings, someone was found in the total quiet room three times and in the partial quiet area five times. The low utilization factor, however, should not be taken as an argument against such areas. It may well be a necessity to provide these spots in order to made the nonterritorial concept acceptable.

TOTAL FLOOR SPACE
Product engineers spend a large proportion of their time outside their own office area; they visit the production line frequently and coordinate their work closely with people in other parts of the plant. In addition, there are the usual meetings that must be attended outside the area, training courses, and absences from work.

An examination of the proportion of the department actually present in the facility at any given time produces some interesting possibilities (figure 9.8). The median proportion present during any sampling is 62.5 percent. The distribution is very skewed, however, so that there is a reasonably high probability of finding as many as 80 percent of the department in that area. Only about 11 percent of the time is there more than 80 percent of the department present in the area. The amount of floor space allocated to the department could thus be reduced by as much as 20 percent (or the number of people assigned to the areas increased by 25 percent) with little danger of overcrowding. The proportion of people present in the area appears independent of the number of people assigned to the department, at least within the range of thirteen to seventeen.

Under the territorial approach, whether closed office or open plan, a certain number of square feet must be assigned to each individual, and when he is absent, it must remain unutilized. With the nonterritorial scheme, an individual is not assigned a specific area of so many square feet but is allowed the same amount of area (or more) with no specification as to location. An area will, therefore, go completely unoccupied far less often. As one person moves out, another moves in. By the time the first returns, someone else will have left, and so on. This will be a very important consideration in many cases in which the actual utilization factor can fall far below the 80 percent found in the experimental department.

INTERDEPARTMENTAL COMMUNICATION

Communication with other departments was measured to determine whether, by increasing the cohesiveness and degree of communication within the department, the new facility caused the department to isolate itself more from the rest of the organization. There is a substantial body of evidence to show that as groups increase in cohesiveness, they tend more and more to seal themselves off from external contact and influences (see Pelz and Andrews, 1966, chap. 13). There was a fear, therefore, that by enhancing the group's internal coordination and identity, the new facility might detract from external communication. It was quite surprising to find that for a short period of time following the facility conversion, there was actually an increase in the level of interdepartmental communication. This led to the tentative decision in December that the nonterritorial facility may have improved the degree of contact with other departments. Over the long term, however, communication dropped back to its old level (figure 9.9). The temporary increase was most probably due to curiosity, which attracted people from other departments into the new facility. After four or five months, the novelty wore off, and fewer people were drawn in. Interaction with other departments had returned to its prechange level. The dashed lines in figure 9.9

Figure 9.9 Interdepartmental Communication Before and After the Introduction of the Nonterritorial Office

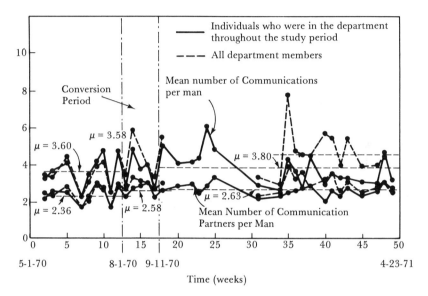

Time (weeks)

show the level of interdepartmental communication when all members of the department (not just those who were present throughout the study) are included in the analysis. Up to the point of facility conversion, this had little effect upon the mean interaction level. After conversion, however, as new engineers were brought in, they retained contact with members of their old group for some period of time and thereby increased the level of interdepartmental communication. By completion of the study, the oldest transfer had been with the experimental department a little over four months, and his communication behavior was approaching that of the other department members. New transfers probably communicated more outside the department for the first six months or so following their transfer, but then they tended to behave like any other member of the department. The continuous turnover of personnel between departments has the very beneficial side effect of promoting interdepartmental communication and preventing the isolation of departments from concern with the rest of the organization (compare chapter 7 of this book).

Communication outside the plant was unaffected by the facilities change. This is to be expected. The measurement was made only as an additional check on any "experimental effect." If the group members were inclined to overreport communication within the plant, they might conceivably, overreport external contacts as well. The fact that they did not lends greater credibility to the internal communication measurements.

PERFORMANCE OF THE DEPARTMENT

The strongest statement that can be made about the department's performance is that it has not changed as a result of the introduction of a nonterritorial office. Performance was measured through structured interviews with members of other departments in the company who served as internal "customers" to the experimental department. Eleven individuals were interviewed in June 1970 (before the facility change). Of these only four felt they still had sufficient contact with the department to evaluate it in May 1971. For this reason, a person from the same department was substituted for one of the June 1970 evaluators, and ratings obtained in December 1970 were used for four more. This gave a total of nine separate before and after evaluations but with no control over individual differences among the evaluators. There is no way to know what the other seven would have said had they remained in contact with the department. The use of substitute evaluators obviously provides no before and after control because of the tendency for individuals to generally be either easy or harsh in their evaluations. Any concern over this should disappear, however, once the results are viewed. There is no apparent difference between 1970 and 1971 in the department's performance. While the general performance measure decreased slightly but not significantly, performance along specific dimensions showed a nonsignificant increase (table 9.2).

It is hardly necessary to enumerate all of the possible reasons for the failure to detect any performance differential. The most obvious is the loss of seven out of eleven evaluators. But it is entirely possible that the improved communication did not and will not improve the department's performance. On the other hand, eight months may be an insufficient period of time for the effects of better communication to be realized. Finally (and perhaps this is a very important consideration), there is every reason to believe that at least some of the evaluators were jealous of all of the attention being received by the experimental department and this jealousy influenced their evaluations. This possibility was suggested during the course of the evaluation interviews when many disparaging

Table 9.2 Departmental Performance Before and After the Nonterritorial Office

Performance Measure	Before	After	p
General appraisal	3.75	3.52	0.22
Aggregation of evaluations along four dimensions of performance	2.66	3.11	0.15

remarks were made, such as, "We don't need such a fancy work area over here. We produce anyway." It is extremely difficult to determine which of the explanations holds the greatest weight. However, after listening to the "sour grapes" expressed by some of the evaluators, it is surprising that their judgment of performance did not decrease.

CONCLUSIONS FROM THE EXPERIMENT

The most important and most obvious conclusion to be drawn from the experiment is that the nonterritorial idea works. It not only reduces facilities costs by eliminating the need for rearranging the physical space every time an area is reorganized (Donofrio, 1970), but it also allows for the allocation of space based upon an expected population density at any point in time. More important than cost savings, however, is the fact that people find it comfortable to work in. Those who have experienced it prefer the nonterritorial area over traditional forms of office arrangement. Furthermore, communication and coordination with the experimental department have increased significantly. And while it is not reflected in the data, this increase will undoubtedly help to improve the department's performance in the long run.

The nonterritorial concept has now proven successful with product engineers and can be readily adapted to similar groups. It is most likely to succeed with groups that spend high proportions of their time outside their office area. Groups who spend most of their time in other areas are accustomed to moving around on the job and are more likely to accept the loss of a permanent individual station. It is with such groups, too, that the most return can be gained from designing for expected levels of occupancy.

It is quite easy, at this point, to envision situations in which entire production lines could be surrounded by nonterritorial areas. Each function—product engineering, quality control, and so on—would have an area marked off by carpet color, partial walls, and other partial visual blocks and would be allowed positions only within the reach of its switchboard. There would be improved coordination within functions, and due to the absence of full walls, access and communication among the functions would be freer. There are many situations in which such a layout is a feasible possibility. The potential for cost reduction and improved coordination around the production line is certainly great enough to warrant further experimentation and adaptation.

A remaining key question is that of introducing the idea to new occupants. When first suggested, it produces, at best, mixed reactions. In fact, it can provoke a good deal of fear or even panic among those who have not yet experienced it. Figure 9.5 demonstrates the effect of experience in reducing this fear. If employees thought that nonterritoriality were going to be adopted widely through an organization, it is easy to imagine the panic it might cause. Some means must be sought to produce the experience without inducing the fear.

Fortunately there is a solution. Every large organization makes frequent use of temporary teams, such as task forces and proposal teams. Moreover, there is usually some difficulty in locating space suitable for these teams. The nonterritorial office, with its inherent flexibility, is natural for such use. An area could easily be set aside for the use of temporary teams and laid out in a nonterritorial fashion. This would pose no threat to team members since the situation would be a temporary one and they would return eventually to the security of their old offices. During their exposure to the facility, however, they may very well come to like it as much as the experimental group in the present study did, and a grassroots demand could then develop. In the meantime, there is the added bonus of improved coordination within the temporary team.

Additional experimentation is certainly necessary. It must be determined just where the nonterritorial office will and will not work. It must, therefore, be tried in functions other than product engineering and on a small quasiexperimental scale in the beginning. Any widespread use must be carefully planned in its introduction. After considerable thought, we believe that the approach of using temporary teams is the best one. This will be true at least until a sufficient number of people have had experience with the innovation, in order to minimize the fear reaction.

NOTES

1. Improved communication will certainly lead to improved performance in many other activities as well. Allen (1964, 1966b, 1969), Pelz and Andrews (1966), Baker, Siegman, and Rubenstein (1967), Shilling and Bernard (1964), and Parker, Lingwood, and Paisley (1968) have all shown performance to be strongly correlated with communication for research engineers and scientists.

2. This idea was first conceived by Armand Beliveau of IBM. It was he who originated the concept of a nonterritorial office, and his management perceived the need for experimentation (see Donofrio, 1970).

3. Analysis of variance with nested classification.

10 WHERE ARE WE?

SCIENTIFIC AND TECHNOLOGICAL INFORMATION POLICY
Since concerns over the information explosion in science and technology first began to emerge some twenty years ago, attempts at resolution have focused on but one small part of the total system. Enormous efforts have gone into improving the preparation, delivery, and accessibility of technical documentation. Governments, private firms, and professional societies are making very extensive investments in the technical information areas. Such efforts have universally emphasized formal information services such as library expansion and interlibrary cooperation in acquisition of publications, bibliography and indexing, abstracting information analysis, translations, reviews, document reproduction, and computer-based access and retrieval. Informal person-to-person communication has been recognized but not truly grappled with or taken into account in planning overall systems. As a typical example, UNESCO, in its comprehensive feasibility report, "UNISIST—A World Science Information System," recognizes three broad categories of information source: informal, person-to-person relations; formal sources including published literature and unpublished documentation; and sources of raw data, usually tabular in form. In discussing the first of these they acknowledge the importance of person-to-person communication as demonstrated in several research studies. The UNESCO report then goes on to devote most of the rest of its attention to the two formal categories. In devising a "world science information system or network," they confine its scope to "scientific and technical information" as embodied mainly in formal, printed sources. The role of the scientists themselves in the proposed system is limited to assisting in the preparation of material for input to the system through indexing, abstracting, and so on, and at the other end, by learning more about how to use the formal system efficiently. This approach is quite typical of that taken by both national and supranational bodies, as well as the management of many firms.

No one at any of these organizational levels seems willing to take on the difficult problem of dealing with the engineers and scientists themselves as communicators. This is true despite the general acknowledgment of their importance.

The principal purpose of the present book is to demonstrate that something can be done to control and improve the functioning of this important mode of communication. The problem is not an impossible one. All of the answers cannot be given here. What has been provided is an ample demonstration that sound analysis can lead to better understanding of the problem and a modest beginning in the direction of potential improvements. I hope that this demonstration will encourage others to further develop this important area of research.

ORGANIZATIONAL BEHAVIOR

The study of organizational behavior has long skirted the subject of communication behavior, with very few attempts at more than speculation. Guetzkow (1965) points out that up to the time of his review, the empirical investigation of communication in an organizational setting had been almost totally neglected. Roberts and her colleagues (1974) further substantiate this claim and show that little has changed in the intervening nine years. They conclude, further, that although theorists almost unanimously acknowledge the importance of communication as a construct for the study of organizations, there is virtually no agreement of definitions or of the way in which the construct should be employed. On the other hand, they do see fertile ground for research:

It is clear that there exists no shortage of researchable variables in this area, and some of the kinds of questions which might be asked from each theoretical vantage point are suggested. There also exist no theoretical models for integrating organizational and communication variables. Further, communication theorists pay little attention to organizations and organizational theorists little attention to communication.

In retrospect, we find what might be termed a failure to communicate. The diverse use of the term communication by organizational analysts merely provides suggestions as to the ultimate form research and theory might take. We need, it seems, at least three approaches to the problem:

1. Specific definitions of various facets of organizational communication which can be operationalized and investigated across levels.
2. Integration of the implications of the various organizational theoretical positions in terms of communication, which will, in turn, suggest
3. Research programs to describe organizational communication and to predict the relationships of its various facets to other organizational variables.

The present research is a step in the direction of initiating the third approach. Certainly the research methods used in this work can be applied to nontechnical areas as well.

In conclusion, I hope that the results of the research will influence three bodies of people. First are those concerned with science policy, particularly its information aspects. This research should encourage them to broaden their concerns to include interpersonal communication as well as documentation. I have demonstrated that there is a potential for policy development in the area of improving direct contact among scientists and technologists. Second are the managers of research and development organizations. With them, I stress the importance of communication both within and outside their organizations. I have provided them with some tools to improve the performance of their organizations. Finally are the students of organizational behavior. I hope that I have been able to demonstrate to them the use of some new tools for the analysis and study of organizations.

APPENDIX
INSTRUMENTS USED IN DATA COLLECTION

THE TIME ALLOCATION FORM

This form was used on some of the projects studied. It was completed at the end of each day by the cooperating engineer. At the end of the week, it was returned. Each week a new copy was mailed to the respondent with his solution development record. On the reverse side was an additional form to record the use of library and other written material on a daily basis.

TIME ALLOCATION FORM

Name _____ Date_____

Day	Sun.	Mon.	Tues.	Wed.	Thurs.	Fri.	Sat.
Activity		Date					
Analytic Design							
Literature Search							
Consultation with Specialists within the Company (but not on the research team)							
Consultation with Experts outside of the Company (paid or unpaid, formal or informal)							
Total Time on the Problem							

LITERATURE SOURCES

Please list the texts, books, periodicals, internal company reports, etc., which you referred to in your literature search, the function which the information served or was intended to serve, and an estimate of the approximate time which you spent with each source. As a publication is used, log it on this form; at the end of the three week period, estimate the total time for each source.

The following code should be used to indicate the function for which the information was used or was to be put to use, and how the information was obtained:

Acquisition Code
A. Personal library search
B. Discovered through use of technical abstract
C. Search by library assistant
D. On desk or personal file
E. Borrowed from colleague

Function Code
1. Aid in direct solution of a problem (e.g., handbook-type information)
2. Determination of results of related work performed by others
3. Determination of procedures (e.g., testing procedures, fabrication procedures)
4. Learning new specialty—broadening areas of attention (to include brushing up on an old specialty)
5. Browsing of technical literature, resulting in significant discovery
6. Verifying reliability of an answer
7. Keeping abreast of developments in one's own particular field
8. Keeping abreast of developments on related or competing systems
9. Aid in definition of operational environment

Title of Journal, or Book (Indicate if internal company document)	Method of Acquisition (by code no.)	Function (by code number)	Total time (in hours)

THE SOLUTION DEVELOPMENT RECORD

The solution development record necessarily evolved over time. The version shown in chapter 2 was supplemented later with an additional section to provide more complete understanding of the reasons why certain probabilities were encircled.

In this version, the respondent, in addition to estimating probability, is also asked to indicate for each alternative his certainty over the likelihood that it will satisfactorily perform its intended mission. This dimension is quite independent of probability and is best described by considering situations in which a probability of 0.5 is assigned to each of two alternatives. This could result from either of two situations. First could be a case in which both alternatives are equally satisfactory and it makes little difference which is chosen: either one will work. A quite different circumstance arises when both alternatives are equally unsatisfactory but one must be chosen since nothing else presents itself. The engineer or scientist's information-gathering behavior would be expected to differ depending upon which of the two situations he faced. The additional data often enabled a more complete understanding of the reasons underlying observed behavior.

SOLUTION DEVELOPMENT RECORD
MANNED URANUS LANDING IN AN EARLY TIME PERIOD STUDY
GENERAL UNITED AEROSPACE CORPORATION
Name _____   Date _____
Subproblem #1: Design of
the electrical power supply
subsystem for the space   Estimate of Probability that Alternative Will Be
vehicle   Employed
a. Alternative approaches:
   hydrogen-oxygen fuel cell   0 0.1 0.2 0.3 0.4 0.5 0.6 0.7 0.8 0.9 1.0

   KOH fuel cell   0 0.1 0.2 0.3 0.4 0.5 0.6 0.7 0.8 0.9 1.0

   Rankine cycle thermal re-
   actor   0 0.1 0.2 0.3 0.4 0.5 0.6 0.7 0.8 0.9 1.0

   Brayton cycle reactor   0 0.1 0.2 0.3 0.4 0.5 0.6 0.7 0.8 0.9 1.0

   _____   0 0.1 0.2 0.3 0.4 0.5 0.6 0.7 0.8 0.9 1.0

   _____   0 0.1 0.2 0.3 0.4 0.5 0.6 0.7 0.8 0.9 1.0

b. Alternative approaches: Estimate of the Likelihood that Alternative Will Satisfactorily Perform (meet or exceed specification)

hydrogen-oxygen fuel cell   certain   very   rather   tossup   somewhat   quite
                                      likely likely   unlikely   un-
                                                                           likely

KOH fuel cell   certain   very   rather   tossup   somewhat   quite
                                      likely likely   unlikely   un-
                                                                           likely

Rankine cycle thermal re-   certain   very   rather   tossup   somewhat   quite
actor   likely likely   unlikely   un-
                                                                           likely

Brayton cycle reactor   certain   very   rather   tossup   somewhat   quite
                                      likely likely   unlikely   un-
                                                                           likely

_____   certain   very   rather   tossup   somewhat   quite
                                      likely likely   unlikely   un-
                                                                           likely

_____   certain   very   rather   tossup   somewhat   quite
                                      likely likely   unlikely   un-
                                                                           likely

c. If information which had a serious impact upon your visualization of the problem or any of its potential solutions was received at any time during the past week, please circle the source(s) of that information on the line below. Sources are defined on the reverse side.

Information Source: L   V   C   ES   TS   CR   PE   E

comments (if any): _____

d. Please estimate the percentage technical completion of the portion of the project with which you are concerned____%. (This need not be a monotonically increasing function of time. Since it is a subjective estimate and since R&D is characterized by the continual discovery of new problem areas, this estimate may decrease as well as increase from week to week),

THE COMMUNICATION SURVEY QUESTIONNAIRE

There were many versions of this questionnaire. Each was tailored to the organization at hand. The ones shown here are fairly typical of those used, however. There are two separate forms: one gathers background data on each individual, and the other samples communication on a weekly basis over some period of time.

SCIENTIFIC/TECHNOLOGICAL COMMUNICATION SURVEY

1. Name _____ Date _____

2. Department No. _____. Group No. _____.

3. Please indicate the building grid number (for example, C-3) on the pillar nearest your desk _____ .

4. Your age _____ .

5. Number of years of technical experience in specific field in which you are currently engaged: _____ yrs.

6. Since receipt of your last degree, have you taken any additional scientific, engineering, or math courses?
   a. yes _____ no _____ b. how many? _____ c. number of years since last course _____ yrs.

7. How many years have you been with _____? _____ yrs.

8. Please indicate the following:
   a. Date of baccalaureate degree _____ (year).
      Please circle one: B.S. B.A. In what field _____
   b. Advanced degrees
      M.S. date _____ field _____
      M.A. date _____ field _____
      Ph.D. or Sc.D. date _____ field _____

9. To what professional societies do you belong?
   IEEE AAAS AIME AAS ASME Amer. ACS Amer. AIAA Amer.
   Inst. Math. Phys.
   Biol. Soc. Soc.
   Sc.
   AMA ORSA other(s) _____
   (specify)

10. In the past year, how many meetings sponsored by professional societies did you attend?
    0 1 2 3 4 5 or more

11. During your professional career, how many papers have you presented at professional society meetings?
    0 1 2 3 4 5 6 7 8 9 10 11 or more

12. How many in the past year?

    0   1   2   3   4   5 or more

13. How many patents do you hold (include those pending)?

    0   1   2   3   4   5   6   If more than 6, how many? _____

14. Please list the titles and journals of publication for any papers that you have published since joining your present group.

Title	Journal	No. of co-authors

15. Do you attend the meetings of local chapters of professional societies (e.g., Los Angeles section of IEEE)?

Yes _____ No _____

16. If yes, what societies?

society	every month	every other month	every six months	once a year	less often
IEEE					
AIME					
ASME					
AIAA					
ACS					
Amer. Inst. Biol. Sc.					
Amer. Math. Soc.					
AMA					
ORSA					
AAS					
Other _____					

          (specify)

17. Please name those people in the Advanced Systems & Technology organization with whom you most frequently discuss technical matters (once a week or more).

18. Please name the individuals from this organization with whom you most frequently eat lunch

How frequently?

Name	daily	twice/week	once/week	less

19. Do you consistently (twice a year or more) meet with any *other* (outside of Advanced Systems & Technology) technically trained individual or individuals to eat lunch or dinner?

Yes _____ No _____

20. If yes, please indicate the institutional affiliations of these people.
    How frequently?

	twice a year	bimonthly	monthly	more often
_____	_____	_____	_____	___
_____	_____	_____	_____	___

21. If you had a new idea for a research project, to whom, in the Advanced Systems & Technology Organization, would you first express this idea?

    _____

22. When you encounter a particularly "hairy" technical problem (in your own technical specialty), please indicate the names of anyone within the organization to whom you would turn for assistance.

    _____   _____
    _____   _____

23. Anyone outside the organization?
    Yes _____   No _____

24. Please name *his* organization. _____

25. Think back to your last completed research project. Try to identify the most difficult technical obstacle or subproblem that you had to resolve in the course of this job. Would you please indicate the sources of information which were especially helpful in overcoming this obstacle?

    _____ attending papers at conventions
    _____ attending symposia at conventions
    _____ scanning or reading of journals
    _____ informal discussion at conventions
    _____ preprints, reprints, or abstracts from author
    _____ books or monographs
    _____ informal discussion with colleagues within Advanced Systems & Technology
            who? _____
    _____ informal discussion with colleagues outside the Advanced Systems & Technology organization
            Please indicate the organizational affiliation of these colleagues.

    _____

    _____ verbal or written reports from assistants
    _____ other

26. How long have you reported to your present supervisor? _____ months

27. Please name any people in your present work group (section) with whom you have worked before (in a different group).

name	duration of earlier association (months)
_____	_____
_____	_____

28. Please name those people from the Advanced Systems & Technology organization that you meet most frequently on social occasions (in the evening, weekends, etc.).

_____   _____

_____   _____

29. During the past five years have you been invited to participate in any government-sponsored, or professional society-sponsored ad-hoc groups (for example, Defense Science Board Task Forces)?

Yes _____ No _____

If yes, how many?

1   2   3   4   5   6

30. Please indicate the number of proposal teams in which you participated during the past year.

0   1   2   3   4   5   6   7   8

31. Please indicate (by a check mark) which of the following periodicals you subscribe to, and which you read regularly.

periodical	subscribe	read regularly	periodical	subscribe	read regularly
Proc. IEEE	____	____	Journal of Operations Research Society	____	____
Physics Today	____	____	Computers and Automation	____	____
Scientific American	____	____	Datamation	____	____
Mechanical Engineering	____	____	Industrial Research	____	____
Chemical & Engineering News	____	____	International Science and Technology	____	____
Civil Engineering	____	____	Product Engineering	____	____
American Engineer	____	____	Microwave Journal	____	____
Electronic Design	____	____	Design News	____	____
American Scientist	____	____	Acoustical Society Journal	____	____
Electrical Engineering	____	____	Electronic Industries	____	____
Electronics	____	____	Electro-Technology	____	____
Nucleonics	____	____	Current Contents	____	____
Aerospace Engineering	____	____	Chemical Abstracts	____	____
Science	____	____	Journal of Metals	____	____
Space-Aeronautics	____	____	Journal of the ECS	____	____
Astronautics and Aeronautics	____	____	Electrochemical Technology	____	____
Missile Design & Development	____	____	Chemical Engineering News	____	____
Journal of Applied Physics	____	____	Journal of Physical Chemistry	____	____
Metals Progress	____	____			

Metals Review	____ ____	Physical Abstracts	____ ____
Any of the IEEE		Physical Review	____ ____
Transactions (if	____ ____	AIAA Journal	____ ____
checked, please		Control Engineering	____ ____
indicate, by title,		IEEE Spectrum	____ ____
which transac-		Nature	____ ____
tions):	_____	Space & Planetary	
	_____	Sciences	____ ____
	_____	Astronautical Jour-	
	_____	nal	____ ____
	_____	Astronomical Jour-	
	_____	nal	____ ____

32. What other professional, technical, or scientific journals (excluding trade
publications) do you subscribe to, or read regularly? (please check)

                                       subscribe   read regularly

_____   ____   ____

_____   ____   ____

On the average, how many hours/week do you spend reading the above
literature?____

Thank you!

COMMUNICATION STUDY (TIME SERIES)

Name _____

Location _____ Date _____

PLEASE NOTE THE DATE STAMPED ABOVE. PLEASE COMPLETE THIS QUESTIONNAIRE ON THIS DATE AS CLOSE TO THE END OF YOUR WORK DAY AS POSSIBLE. Your cooperation is sincerely appreciated. Please indicate if you have communicated with any of the individuals listed during *this* day and whether it was *you* or *he* who initiated the communication by checking the appropriate place. Please consider only those communications which were concerned with technical matters. Please return to S. Cooney to be forwarded for analysis to Professor T. Allen.

	face-to-face initiator		telephone initiator	
HEADQUARTERS	you	he	you	he
J. Armstrong				
W. Barry				
P. Broughan				
M. Butler				
F. Cantwell				
D. Conniffe				
S. Cooney				
A. Cox				
O. Daly				
O. G. Daly				
A. J. Fitzgerald				
P. V. Geoghegan				
B. Gilsenan				
S. Gilroy				
J. D. Golden				
T. Higgins				
M. F. Keane				
J. Kilroy				
B. Lewis				
P. Markey				
P. C. Mulleady				
E. McCormick				
P. M. McDonnell				
M. Neary				
L. Noone				
I. M. O'Deirg				
A. P. O'Reilly				
M. O'Sullivan				
E. Pitts				
V. Reilly				
D. Richardson				
P. Ryan				

C. Shouldice

J. J. Sugrue

T. Walsh

E. Wymes

CREAGH

P. J. Daly

J. J. Fitzgerald

M. J. Hope-Cawdery

A. Kearney

T. Nolan

W. Sheehan

J. Sreenan

B. W. H. Stronach

V. M. Timon

REFERENCES

Abrams, R. A. 1943. Residential propinquity as a factor in marriage selections. *American Sociological Review* 8: 288–294.

Allen, T. J. 1964. The use of information channels in R&D proposal preparation. Cambridge, Mass.: M.I.T. Sloan School of Management, Working Paper No. 97-64.

_____1966a. Managing the flow of scientific and technological information. Ph.D. dissertation, M.I.T. Sloan School of Management.

_____1966b. Performance of information channels in the transfer of technology. *Industrial Management Review* 8: 87–98.

_____1969. Communication networks in R&D laboratories. *R&D Management* 1: 14–21.

_____1970. Communication networks in R&D laboratories. *R&D Management* 1: 14–21.

Allen, T. J., and Frischmuth. 1969. A model for the description and evaluation of technical problem solving. *IEEE Transactions on Engineering Management* 16: 58–64.

Allen, T. J., and Marquis, D. G. 1963. Positive and negative biasing sets: The effect of prior experience on research performance. *IEEE Transaction on Engineering Management* 11: 158–162.

Andrews, F. M., and Farris, G. F. 1967. Supervisory practices and innovation in scientific teams. *Personnel Psychology* 20: 497–515.

Baker, N. R., Siegmann, J., and Rubenstein, A. H. 1967. The effects of perceived needs and means on the generation of ideas for industrial research and development projects. *IEEE Transactions on Engineering Management* 14: 156–162.

Bar-Hillel, Y., and Carnap, R. 1953. Semantic information. *British Journal of the Philosophy of Science* 4: 147–157.

Battelle Memorial Institute. 1973. *Interactions of science and technology in the innovation process: Some case studies.* Final report to the National Science Foundation NSF-C667, Columbus, Ohio.

Berul, L. H., Elling, M. E., Shafritz, A. B., and Sieber, H. 1965. *DOD user needs study.* Philadelphia: Auerbach Corporation.

Beveridge, W. I. B. 1957. *The Art of Scientific Investigation.* New York: Morton.

Blau, P. M. 1963. *The Dynamics of Bureaucracy.* Chicago: University of Chicago Press.

Burton, R. E. 1959a. Citations in American engineering journals I. Chemical engineering. *American Documentation* 10: 70–71.

_____ 1959b. Citations in American engineering journals III. Metallurgical engineering. *American Documentation* 10: 135–137.

_____ 1959c. Citations in American engineering journals III. Metallurgical engineering. *American Documentation* 10: 209-213.

Carnap, R. 1962. *Logical Foundations of Probability.* 2d ed. Chicago: University of Chicago.

Cherry, C. 1957. *On Human Communication.* Cambridge, Mass.: The MIT Press.

Cohen, A. R. 1958. Upward communication in experimentally created hierarchies. *Human Relations* 11: 41-51.

Coleman, J., Katz, E., and Menzel, H. 1966. *Medical Innovation: A Diffusion Process.* New York: Bobbs Merrill.

Donofrio, A. M. 1970. When the walls came tumbling down. *IBM Magazine* 2: (no. 11), 21-23.

Feller, W. 1950. *An Introduction to Probability Theory and Its Applications.* New York: Wiley.

Festinger, L., Schacter, D., and Back, K. 1950. *Social Pressures in Informal Groups: A Study of Human Factors in Housing.* New York: Harper.

Fishendon, R. M. 1959. Methods by which research workers find information. *Proceedings of the International Conference on Scientific Information.* Washington: National Academy of Science.

Flament, C. 1963. *Applications of Graph Theory to Group Structure.* New York: Prentice-Hall.

Frohman, A. 1968. Polaroid library usage study. Cambridge, Mass.: M.I.T. Sloan School of Management, term paper.

Frost, P., and Whitley, R. 1971. Communication patterns in a research laboratory. *R&D Management* 1: 71-79.

Fusfeld, A. R., and Allen, T. J. 1974. *Optimal height for a research laboratory.* Cambridge, Mass.: M.I.T. Sloan School of Management, Working Paper No. 699-74.

Galbraith, J. 1969. *Organization design: An information processing view.* Cambridge, Mass.: M.I.T. Sloan School of Management, Working Paper No. 425-969.

Gerstenfeld, A. 1967. Technical interaction among engineers and its relation to performance. Ph.D. dissertation, M.I.T. Sloan School of Management.

Gibbons, M., and Johnston, R. D. 1974. The roles of science in technological innovation. *Research Policy* 3: 220-242.

Gilnisky, A. S. 1951. Perceived size and distance in visual space. *Psychological Review* 57: 460-482.

Goffman, E. 1956. *The Presentation of Self in Everyday Life.* New York: Anchor Books.

Guetzkow, H. 1965. Communications in organizations. In *Handbook of Organizations,* ed. J. March, 534–573. Chicago: Rand McNally.

Gullahorn, J. T. 1952. Distance and friendship as factors in the gross interaction matrix. *Sociometry* 15: 123–134.

Hagstrom, W. 1965. *The Scientific Community.* New York: Basic Books.

Harary, F., Norman, R., and Cartwright, D. 1965. *Structural Models: An Introduction to Graph Theory.* New York: Wiley.

Hare, P. H., and Bales, R. F. 1963. Seating position and small group interaction. *Sociometry* 26: 480–486.

Helms, J. D. 1970. Communication between diverse technical groups. Master's thesis, M.I.T. Sloan School of Management.

Henize, J. 1968. The use of informal reports in R&D. Cambridge, Mass.: M.I.T. Sloan School of Management, term paper.

Herner, S. 1954. Information-gathering habits of workers in pure and applied science. *Industrial Engineering Chemistry* 46: 228–236.

Homans, G. C. 1961. *The Human Group.* New York: Harcourt.

Hurwitz, J., Zander, A., and Hymovitch, B. 1960. Some effects of power on the relations among group members. In D. Cartwright and A. Zander, eds., *Group Dynamics,* 2d ed. New York: Harper.

IIT Research Institute. 1968. *Technology in retrospect and critical events in science.* Report to the National Science Foundation NSF C-235.

Kanno, M. 1968. Effects in communication between labs and plants of the transfer of R&D personnel. Master's thesis, M.I.T. Sloan School of Management.

Kaplan, N. 1965. *Science and Society.* New York: Rand McNally.

Katz, D., and Kahn, R. L. 1965. *The Social Psychology of Organizations.* New York: Wiley.

Katz, E., and Lazarsfeld, P. F. 1955. *Personal Influence: The Part Played by People in User Communication.* Glencoe: Free Press.

Kelley, H. H. 1951. Communication in experimentally created hierarchies. *Human Relations* 4: 39–56.

Kennedy, R. 1943. Premarital residential propinquity. *American Journal of Sociology* 18: 580–584.

Krulee, G. K., and Nadler, E. B. 1960. Studies of education for science and engineering: Student values and curriculum choice. *IEEE Transactions on Engineering Management* 7: 146–158.

Kuhn, T. B. 1970. *Structure of Scientific Revolutions.* Rev. ed. Chicago: University of Chicago Press.

Kunnapas, T. M. 1958. Measurements of subjective length in the vertical-horizontal illusion. *Nordisk Pskologic* 10: (no. 4-5) 203–206.

Langrish. J. 1971. Technology transfer: Some British data. *R&D Management* 1: 133-136.

Lazarsfeld, P. F., Berelson, B., and Gaudet, H. 1948. *The People's Choice.* New York: Free Press.

Leavitt, H. J. 1951. Some effects of certain communication patterns on group performance. *Journal of Abnormal and Social Psychology.* 46: 38-50.

Maisonneuve, J. 1952. Selective choices and propinquity. *Sociometry* 15: 123-134.

Marquis, D. G., and Allen, T. J. 1966. Communication patterns in applied technology. *American Psychologist* 21: 1052-1060.

Marquis, D. G., and Straight, D. L. 1965. Organizational factors in project performance. Cambridge, Mass.: M.I.T. Sloan School of Management, Working Paper No. 133-165.

Mednick, S. A. 1962. The associative basis of the creative process. *Psychological Review* 69: 220-232.

Menzel, H. 1960. *Review of studies in the flow of information among scientists.* New York: Columbia University, Bureau of Applied Social Research.

Morton, J. A. 1965. From physics to function. *IEEE Spectrum* 2: 62-64.

Newcomb, T. M. 1961. *The Acquaintance Process.* New York: Holt.

Osborne, A. F. 1957. *Applied Imagination.* New York: Scribner.

Parker, E. B., Lingwood, D. A., and Paisley, W. J. 1968. *Communication and research productivity in an interdisciplinary behavioral science research area.* Stanford: Institute for Communication Research, Stanford University.

Pelz, D. C., and Andrews, F. M. 1966. *Scientists in Organizations.* New York: Wiley.

Peters, J. R. 1970. A case study of research and development communication in a small chemical firm. Master's thesis, M.I.T. Sloan School of Management.

Price, D. J. DeSolla. 1965a. Networks of scientific papers. *Science* 149: 510-515.

_____ 1965b. Is technology independent of science? *Technology and Culture* 6: 553-568.

_____ 1970. In D. K. Pollock and Nelson, C. E. (eds.) *Communication Among Scientists and Technologists.* Lexington, Mass.: Heath.

Ritti, R. R. 1971. *The Engineer in the Industrial Corporation.* New York: Columbia University Press.

Roberts, E. B., and Wainer, H. A. 1971. Some characteristics of technical entrepreneurs. *IEEE Transactions on Engineering Management,* EM-18, 3.

Roberts, K. H., O'Reilly, C. A., III, Bretton, G. E., and Porter, L. H. Organizational theory and organizational communication: A communication failure? *Human Relations* 27: 501-524.

Rogers, E., and Shoemaker, R. 1972. *Communication of Innovations.* New York: Free Press.

Schein, E. H., and Bailyn, L. 1975. Work involvement in technically based careers: A study of M.I.T. alumni at mid-career. M.I.T. Sloan School of Management.

Scott, C., and Wilkens, L. T. 1962. The use of technical literature by industrial technologists. *IRE Transactions on Engineering Management* 9: 78-86.

Shaw, R. R. 1959. Flow of scientific information. *College and Research Libraries* 20: 163-164.

Shapero, A. 1967. Preliminary analysis of inter-specialty mobility of technical professional manpower resources. National Science Foundation.

Shepard, H. A. 1954. The value system of a research group. *American Sociological Review* 19: 456-462.

Sherwin, E. W., and Isenson, R. S. 1967. Project Hindsight. *Science* 156: 1571-1577.

Shilling, C. W., and Bernard, J. W. 1964. Informal communication among Bio Scientists. George Washington University Biological Sciences Communication Project, Report 16A-64.

Sommers, R. 1969. *Personal Space: The Behavioral Basis of Design.* Englewood Cliffs, N.J.: Prentice-Hall.

Steinor, B. 1950. The spatial factor in fact to fact discussion groups. *Journal of Abnormal and Social Psychology* 45: 552-555.

Stodbeck, F. L., and Hook, H. L. 1961. Social dimension of twelve-man jury table. *Sociometry* 24: 397-415.

Taylor, R. 1972. *An analysis of the two-step flow process in a military R&D laboratory.* Unpublished DBA, Indiana University.

Utterback, J. L. 1975. Innovation in industry and the diffusion of technology. *Science* 183: 620-626.

Vincent, R. J., Brown, W., Markley, R., and Arnoult, M. 1968. Magnitude estimation of perceived distance over various distance ranges. *Psychonomic Science* 13: 303-304.

Walsh, V. M., and Baker, A. G. 1972. Project management and communication patterns in industrial research. *R&D Management* 2: 103-109.

Wertheimer, M. 1959. *Productive Thinking.* New York: Harper.

Zipf, G. K. 1949. *Human Behavior and the Principle of Least Effort.* Reading, Mass.: Addison-Wesley.

SUBJECT INDEX

NAME INDEX